REAL LIFE YELLOWSTONE
What Wildlife in America's Premier Park Can Teach Us

SCOTT HERRING

RIVERBEND
PUBLISHING
Essex, Connecticut

RIVERBEND
PUBLISHING
An imprint of The Globe Pequot Publishing Group, Inc.
64 South Main Street
Essex, CT 06426
www.globepequot.com

Copyright © 2026 by Scott Herring

All rights reserved. No part of this book may be reproduced in any form or by any electronic or mechanical means, including information storage and retrieval systems, without written permission from the publisher, except by a reviewer who may quote passages in a review.

British Library Cataloguing in Publication Information available

Library of Congress Cataloging-in-Publication Data

Names: Herring, Scott author
Title: Real life Yellowstone : what wildlife in America's premier park can teach us / Scott Herring.
Description: Essex, Connecticut : Riverbend, [2026] | Includes bibliographical references. | Summary: "A guide to the astounding life-forms of Yellowstone National Park"— Provided by publisher.
Identifiers: LCCN 2025038114 (print) | LCCN 2025038115 (ebook) | ISBN 9781683344667 paperback | ISBN 9781683343752 epub
Subjects: LCSH: Animals—Yellowstone National Park | Human-animal relationships—Yellowstone National Park | Wildlife watching—Yellowstone National Park | Yellowstone National Park | BISAC: NATURE / Regional | TRAVEL / United States / West / Mountain (AZ, CO, ID, MT, NM, NV, UT, WY)
Classification: LCC QL215 .H47 2026 (print) | LCC QL215 (ebook)
LC record available at https://lccn.loc.gov/2025038114
LC ebook record available at https://lccn.loc.gov/2025038115

Contents

Introduction 1

CHAPTER ONE: The Shocking Truth
   The Bears I 10

CHAPTER TWO: Lessons Learned
   The Bears II 45

CHAPTER THREE: By Their Fruits Ye Shall Know Them
   Plants and Pathogens 72

CHAPTER FOUR: One Health, Many Dangers
   Zoonotic Agents and Other Small Wonders 104

CHAPTER FIVE: Nature Red in Tooth and Claw
   The Carnivores 159

CHAPTER SIX: "Man Is in the Forest!"
   The Deer Family 214

CHAPTER SEVEN: A Different Kind of Bullfight
   The Bison 255

Conclusion: Be Not Afraid 291

Introduction

WE NOW LIVE OUR LIVES SURROUNDED BY FAKENESS.

Perhaps not every minute—but surely there has never been a time in history when the human world was so phony and stupid.

The cause—or a big one, anyway—is not far to seek: the glowing screens that vacuum us toward them like a broken window in a pressurized jet airplane. They were tough enough to resist back when the internet began to mature in the late 1990s and into the 2000s, and there were suddenly websites and "web logs" for every imaginable pursuit, serious, semi-serious, and ridiculous. Already, we were developing a large population of men, especially men, who were utterly consumed by massively multiplayer games like Minecraft and Worlds of Warcraft, and never went outside.

Then came the smartphone. The great leap, it is generally agreed, was the introduction of the iPhone by Apple Computer in 2007. At first it was not clear what had happened, and even today, some people—older people especially, like me—are still discovering the full, vast extent of the change. Computers had been big and clunky, made up of a giant *X-Files*-era monitor and tower, giant printer, keyboard, mouse, external speakers, and so on. Getting online required hijacking a landline telephone. We called them "desktops," but they actually filled part of a room, and operating all that stuff required something close to expertise. The iPhone was a computer that anyone could use, and was so portable that young women now use it as what amounts to a fashion accessory, protruding at a flirty angle from a rear hip pocket.

We were all seduced.

What child now grows up outside the influence of all the glowing screens? The statistics are astounding. Nearly all teenagers now have smartphones. A Gallup survey in 2023 found that over half of US teenagers were on social media for at least 4 hours a day, and in that group, girls were on for 5.3 hours, every day. Teens of both sexes were on YouTube for nearly 2 hours, every day, and were on TikTok for an hour and a half. Instagram just barely brought up the rear among those three, at not quite an hour a day. Both sexes spend a good deal more time online than they do on homework.

Girls are on social media more than boys, but boys make up the difference with video games. Another survey, from 2024, arrived at a final total of all online time: It found that kids were looking at screens for over seven hours a day, a number that has increased by two hours just from 2015—and those are average numbers. Some kids, especially those who are shy, awkward, self-conscious about their looks, and not good at sports, are looking at them even longer. It does not help that we are now well into multiple digital generations. If mom and dad were shut-ins when they were younger, they will tend to be more permissive about time spent online by their kids. A 2020 survey by the Mott Children's Hospital at the University of Michigan found that 71 percent of parents believe video games are *good* for teenagers, a number that surely has a great deal to do with wishful thinking.

But that brings us to Yellowstone. It shows us all a way out.

That may seem at first like a leap, but what are we confronted with, when we enter Greater Yellowstone? A place that is in earnest, in a frightening way. There is nothing about it that is fake. You cannot know what it might throw at you, and whatever it does throw may be as real as a hardball coming at you at major-league speed.

Yellowstone, to put it briefly, demands you pay attention. In a place like this, the rules humans normally live by are only one set of rules among many. Leave the roads and the villages, and other rules come into play, because humans are no longer in charge. Other creatures have a say, and they are not always your friends.

Did you know that there are, by my count, four kinds of animals in Greater Yellowstone that will eat people? Did you know that the biggest land animal in North America lives here, and that it is often so ill-tempered that it picks people up and throws them through the air?

Introduction

Did you know that the park hosts the pathogen that gives you bubonic plague? And the thing the news media calls the "brain-eating amoeba"? And probably hantavirus, and probably rabies, and definitely something very close to mad cow disease?

This is Yellowstone—but if you think about it, it is actually unsurprising that a land with such astounding landscapes will also have astounding life-forms.

And that is what makes it real. It is the opposite of AI, and it is mainly the dangerous animals and other life-forms that give it this quality. The terrain is only a threat when you are doing something foolhardy, like wandering around the thermal areas at random or hanging over the edge of a cliff to get a selfie. For a smaller group, this danger comes from getting creative, like climbing a mountain from a route that might turn out to have been a bad idea. Most people, however, never do these things, but just enter the woods, and its creatures are there, even if you do not ever see one. They are there in your imagination, to be sure, but they are also there *somewhere*, maybe right there around that bend in the trail, or behind that stand of trees where a twig just snapped. Their presence is what makes the experience routinely riveting. This is what makes it real.

(And speaking of AI, which is in the news nonstop as I write this—not one sentence in this book was written by AI. Not even a single word.)

The Yellowstone experience does require effort. What we find there today, to switch to the human side of it, is a certain amount of tension between the real and the unreal. Employees in the park used to be utterly cut off from the electronic media produced in the outside world. A television in Mammoth Hot Springs could pull in one or two stations from Billings and Bozeman, depending on how you manipulated the clothes-hanger antenna, but elsewhere in the park, even AM radios did not work. Today, however, the employee housing is wired for internet, and the employees regularly never go outside. The park has cell towers, installed because cell reception provides a way to call for help, but the effect is that people are regularly staring at those screens. Every professional guide in the area has stories to tell about the children of some of their clients, who, when the tour vehicle enters one of the park's many cell shadows, cry, throw a tantrum, and even spin into an anxiety crisis complete with hyperventilation and a sound of panic in the voice. It happens all the time.

Heartening, however, is the fact that a majority never does anything like this, including the kids. As you travel through Yellowstone National Park and hit a traffic jam caused by animals (a "bear jam," "elk jam," etc., in the language of the park), notice how the kids behave, especially the younger kids. They nearly jump through the windows in excitement. They understand that the scene outside those windows is more gripping by far than the scene on the little screens scattered around the interior of the car.

It is the animals that have this effect. Something about the sight of them is so electrifying that all the distractions of the modern world drop by the wayside. They have a power over us still.

What is the nature of this power? It has many facets. For some of the kids, the animals are just cute. For some of the adults, they are just one of the many "sights" of the park. For both, they are another thing to photograph—which, if you have ever tried it, you will understand can be an impossible challenge (the animals just . . . will *not* cooperate). Many

A star and his, or her, adoring fans. GETTY IMAGES

INTRODUCTION

among the visitors are farmers and ranchers, accustomed to sizing up big animals in an appreciative way. Many are hunters, accustomed to the same, if in a more riveting way.

More is happening, though. Look at those big animals long enough, and they begin to remind you of others, just as big, just as awesome. At first, the recollection is hard to place, but it comes in time.

What they look like are the animals in the flickering light at the back of the cave.

We know of quite a few such caves, in various parts of Europe, from the time before humans made it across the land bridge from the Old World. The most famous is in the countryside in southwest France, and is named for a local manor house called Lascaux. In late summer 1940, just as the German air force was beginning to hammer London every night, a dog belonging to a local teenager turned up an unfamiliar hole in the ground. Thinking they had found a secret tunnel leading to Lascaux manor, the teenager and his friends began to explore. In the uncertain light of an oil lamp, they descended into the darkness, until, in that flickering light, they made them out: animals. Hundreds of images of animals painted on the walls, and in the light, they seemed to be moving.

They were painted there by astonishingly talented artists of the Old Stone Age, about seventeen thousand years ago, give or take. They took as their subjects most of the big animals they encountered every day, and that they depended on for every aspect of their being. The walls swarm with food animals, like bison, deer, muskoxen, and horses, and also animals they would have respected and maybe feared, like wolves, lions, and bears. Prominent among the animals are aurochs, the ancestors of today's domestic cattle, although these are not animals you would try to rope and brand. Among them is a figure that art historians call the Great Black Bull, a colossal, almost overwhelming bull aurochs measuring a full seventeen feet long. It dominates the room around it as the actual animal would have dominated the landscapes of ice age Europe.

What did the artists mean by all this? It was certainly not mere doodling or graffiti. The paint did not come from Home Depot; simply acquiring the materials required effort and expertise. The paintings themselves have a boldness about the outlines that suggests the confidence that comes from long practice. Creeping back into this dark place, by the same kind of flickering light that guided its rediscoverers, needed effort and commitment in itself. It required devotion, and cave art, in the end, became a project that went on for generations. It involved a feeling for the animals that can only be called reverence.

What we can see in this cave, and others like it, is an early instance of the fascination and wonder at the life around them that led people everywhere to enshrine animals as either spiritual beings or gods outright. Bulls, creatures of maximal power and virility, were worshipped all over the world, and still are. The Great Black Bull is an early example; a later one (of many) is the bull god Apis of ancient Egypt. The bulls of the bullrings in Spain carry on the tradition. Also worshipped were bears, who, just like the bears in Yellowstone, disappeared into their dens as the world expired in autumn and reappeared with the new life of spring—and so were tailor-made to be the central symbolic figures in the death and rebirth rituals that are a part of religions everywhere.

In Greater Yellowstone you are participating in this ancient tradition—and you are not being pagan by doing so, unless you start making weird sacrifices to the animals. They have the elemental power of the bull in the bullring, and of the old bull gods.

An old piece of advice that rangers give to Yellowstone visitors is important here: They say that to avoid coming to grief in a run-in with a bear, you should hike like an animal, and not like a tourist. What they mean is that safety comes, paradoxically, from escaping the comforts of civilization; as you move down the trail, you need to give your surroundings all your attention, and leave the phone in your backpack. By doing so, you get out from behind the shell of technology that surrounds us all the time and actually experience your surroundings—and that awareness keeps you from blundering into a dangerous animal. You can now actually notice the creature and give it a sensible distance.

And get a thrilling encounter as part of the bargain. When you enter Greater Yellowstone and engage it on its own terms, you become one of its life-forms. You do not have to climb the Grand Teton to have this experience—even just spending the night in one of the campgrounds is enough for a strong taste. When you enter that special space, the animals become like the bulls on the walls at Lascaux. A big part of their power is their dangerousness. It is what keeps our attention at all times, and makes them riveting at moments of high drama.

In this book, we are going to look at those life-forms, and not just animals, but every kind of life that you need to know about to preserve your own—life, that is. My emphasis will be on what they can do to you. Why? Because we all need a taste of the real. If you want to get something worth having from your visit, then you need to give the creatures of Yellowstone all the power they naturally possess. They are even more formidable than you may think.

We will also look at these creatures because they are living the most "real" lives imaginable. There is nothing fake about them or the challenges they present. They are not leading phony lives, and you don't have to, either.

Let them show you how. It is a lesson that follows almost inevitably when you fully enter their world.

We will look at creatures you probably had no idea exist in Greater Yellowstone. Most first-time visitors, for instance, are unaware that there is a species of poisonous snake in the area, and it is not a rarity.

Yellowstone has almost every species of deer native to North America—lacking caribou alone—but only some of them will remind you of Bambi. Because we have so many different kinds of deer, we have the unfriendly ones, too: Be aware that "deer," for a biologist, includes elk and moose. They kick and gore and trample people. A verb you sometimes see when a moose attacks someone is "savage," as in "A visitor was savaged by a bull moose today." It fits. They share precious little in common with Bullwinkle.

We have other grazing animals, and among them is the least friendly of all. The bison, which you will see lumbering around open areas all over the park, is the creature referred to earlier: the biggest land animal in North America. Full-grown males will weigh two thousand pounds and stand six feet high. They can be profoundly ill-tempered. Hardly a summer goes by without multiple "incidents" involving bison and people, because they are also profoundly dangerous. They throw people through the air like a referee's flag in a football game.

We have lions. We have wolves. They mostly leave people alone, although the potential for serious trouble is always there with them. The bears do not leave people alone. We have two kinds, again including nearly all the species in North America (sadly, we lack polar bears). We have the less aggressive black bear—less aggressive, but not harmless, and since they can grow to be over three hundred pounds, they deserve your full respect. Not so their cousin, the grizzly bear. They deserve your fear. It is an open and probably meaningless question what the most dangerous animal in the park actually is, but only one—the grizzly—routinely eats people. Although it happens less often than you may believe, they do seem to think nothing of it.

We will look at all these animals in this book, but will not stop there. Even our plants can hurt you. And more: Most visitors, even experienced ones, are unaware of the amazing range of pathogens in Greater Yellowstone—viruses, bacteria, and parasites that will leave you thinking that they have it in for you.

It is a challenging, demanding place, Yellowstone. It can be terrifying. It all goes together to make the place awesome—the ultimate kind of "real."

You will be safer for having read this book, although people also tend to overdo it. "Safety" is not worth enduring the state of constant fear that some visitors carry with them through the park, from one end to the other until they exit the park. They may relax then, unaware that Greater Yellowstone is enormously larger than Yellowstone National Park. In this book, we will look at the whole place. The Greater Yellowstone Ecosystem is sometimes defined as the present range of the Yellowstone grizzly bear, which is maybe twenty million acres, and probably more. If you are afraid inside the park, you should be afraid until you get all the way to Idaho Falls, Billings, or Salt Lake City. But fear is not necessary. You do not, for instance, need to carry bear spray on the sidewalks at Old Faithful—some people think you do, but that is taking it way too far. Some people are afraid to leave their cars. Yes, the place can be perilous, but the threats can be coped with—and the place would not be the same without them.

INTRODUCTION

So my purpose is not to scare you—although in the pages ahead, you will be scared. Almost no one dies in this book (I will refer you to *Death in Yellowstone*, the classic work by retired Park Historian Lee Whittlesey, where that topic is covered very thoroughly and gruesomely indeed). There will be some exceptions, especially for those who have run afoul of the bears, and a few other varieties of fatal mishap. Fatal or not, the life-forms in this book can be scary. I will show you what they can do, and also show you how you can avoid getting into trouble, which is really not difficult at all.

The reward is enormous: Once you get inside the place, and once you put the phone down, the influence of all fakery over your life is on hold.

I will go so far as to say that I can offer you a vision: In Yellowstone, you can live as bold and red-blooded a life as you wish. It can be the last word in real. I say "can"—it is not guaranteed. But as long as you are heedful of their lessons, and mindful of their demands, the creatures of Yellowstone will show you the way.

CHAPTER ONE

The Shocking Truth

THE BEARS I

SADLY, WE MUST BEGIN WITH A BROKEN PROMISE. YOU MAY RECALL ME saying just now, in the introduction, that no one dies in this book. Notice, though, that it says *almost* no one dies. I hedged, and given the nature of some of the wildlife, it would have been foolish of me not to.

Many of the creatures we will look at are deadly dangerous, or can be. That is what gives them the absorbing, riveting, and potentially terrifying qualities that make them what they are, and that combine to make Yellowstone what it is. As noted, it is an open and unresolvable question just which animals here are most dangerous (although there are only a handful at the top—think of them as future first-ballot inductees to a Hall of Fame of mayhem). Equally unresolvable is the question of which is most frightening, but if we were to run a poll, the winner would surely be the bears.

But which bears? The one you are most likely to see, during a visit to the area, is the American black bear, *Ursus americanus*. It is the smaller of the two bear species in the park, and the common name is accurate: Representatives are usually a glossy black, with a tan muzzle that is often the easiest way to recognize the species. Adult females weigh in the range of 150 to 175 pounds, and adult males around 275, although they can get to be 400. They are generally less aggressive—but not harmless, as we will see later in this chapter. Anyone who has had the increasingly common experience of confronting an agitated pit bull will know that the experience is unnerving at the very least. Bears and bulldogs are equipped with much the same armament: teeth, claws, bite force, speed, and muscle power just in general. It's true the bear is probably less bellicose than a pit bull, but it also *weighs* a great deal more.

Still, the really frightening one is the grizzly, *Ursus arctos horribilis*. There is nothing like them anywhere in the temperate part of North America. They are profoundly dangerous.

And they are unpredictable. You can never be entirely sure that you know what to expect, and the surprises they deliver can be appalling. That unpredictability may be one of the most frightening features of our most frightening animal.

Consider what happened in the early morning hours of July 28, 2010. This incident was as uniquely horrifying as anything in the history of human-bear interactions, and was also uniquely bizarre. These things happen, yes . . . but not like *that*.

It was a typical middle-of-summer night. At Soda Butte Campground, in the Gallatin National Forest, twenty-four of the twenty-seven sites were occupied; in the middle of summer, it should have been full, but compared to the famous places in the region, it was off the beaten track. Unlike some of the campgrounds in Greater Yellowstone, it is a small, obscure, pleasant location. It runs east–west along a thousand-yard stretch of Soda Butte Creek, a mile beyond the little tourist town of Cooke City and five miles outside the Northeast Entrance to Yellowstone National Park. It is thus the responsibility of the US Forest Service. People get the agencies mixed up, and the differences are more important than you may imagine. The National Park Service runs, naturally, the national park, and is a part of the federal Department of the Interior, while the Forest Service is in the Department of Agriculture. Yellowstone National Park is the more heavily visited, crammed with both famous sights and the people seeking them out, and so is more heavily policed. People zero in on the national park, while the national forest is altogether more mellow, so much so that when this one changed its name, people scarcely noticed. It is now the Custer Gallatin National Forest. Two forests were combined into one, and parts of it now stretch all the way to South Dakota.

Mellow or not, this one was ready for all comers: Every campsite had a bear-resistant steel container for storing food—a kind of armored pantry—and there were plenty of bear-proof garbage receptacles. Hanging everywhere were the standard signs warning about the bears. Where food was concerned, they were ready for anything. As was normal, though, an unknown number of campers slept on the ground in tents, with nothing but a wall of nylon as protection against visitors. Oh, and there was bear food in the campground. It was growing all over the place, in the form of what botanists call forbs and most people call herbs, or just flowers. Bears find them delicious, and there are few parts of Greater Yellowstone where they do not grow. It was pretty much inevitable that they would grow here, too.

There is no getting away from it: Yellowstone is more welcoming to the bears than it is to us.

As they are required to do after such an event as took place that night, the many agencies in whose jurisdiction the bears live sent rep-

A Yellowstone grizzly. GETTY IMAGES

resentatives to make up a formal investigative team, along with other experts. The result was an unusually lengthy report (seventy pages—in the past, they were as short as eight). It is as detailed as one could wish for, if one wished for such things, but some details are missing.

For instance, no report on the incident that I can find will tell me exactly how close the people in the campground were, as they slept, to Soda Butte Creek. We know where the campsites were, yes, but that does not necessarily tell us where the people were sleeping, an important detail in a Forest Service campground where the rules are a little looser than in the park itself, just down the road. The creeks in Yellowstone, even the smaller ones, can be surprisingly loud. That sound starts as heavenly mountain music, the kind of sound people come here from all over the world to hear. Eventually, it becomes white noise, and in the end, if—as happens when sleeping on the ground in the high mountains—insomnia develops and a period of hours passes, the water sounds like Nirvana—not the Buddhist state of enlightenment, but the Seattle grunge-rock band (I am entirely serious about that: I once got "Smells Like Teen Spirit" stuck in my head that way). It can easily drown out other sounds. I talked to a botanist who has studied the plants all along Soda Butte Creek, and has stayed in that campground over and over. Willows grow along the creek, a thick shrub-like plant that loves watercourses. The hallmark of that night, as is normal in a bear mauling, was confusion. The vegetation would have further confused matters, since a bear tends to use open areas that border dense vegetation as a way of slipping in and out of sight. Even at night, that can matter. "Sounds wouldn't be completely drowned out by the creek," the botanist concluded. "What *would* happen is they would be masked so much that you would hear the noise, then pull the blanket over your head and go back to sleep."

As Yellowstone goes, it was not unusually dark. Newcomers can be surprised and even frightened by the absoluteness of the darkness in this part of the world, but the moon was only two days past full and high in the sky at that hour, with no rain. Witnesses, and the people directly involved, were understandably vague on the exact time, on the clock, when it all happened, but the official report settled on roughly 2:00 a.m. It was too late for people to be partying. That is a major use of campgrounds in national forests, sadly: They are places where people go to get hammered. We will return to this subject, but for now, this fact of life does help explain how the sequence of events could have unfolded while so many people nearby were unaware. One heard a scream, but dismissed it as a "domestic disturbance," which in the context actually makes perfect sense.

Finally, each campsite is walled in by trees. As the report explained, "Campsites are separated from each other in most cases by visual cover with thick stands of spruce and other conifers providing visual screening." Soda Butte Campground is one with its surroundings in this way; the campground is in a valley, and this part of it is densely forested. It is deeply shaded during the day and, normally, quiet.

The point here is that the human participants were mostly on their own. They were part of the temporary community of the campground, but what happened to them happened in isolation.

And no one noticed, at first, the new arrivals: four figures moving steadily out of the shadows.

What were they doing here? What did they want? No one will ever know, or at least no human. We can only guess. The guess may be educated, but uncertainty is the dominant factor here.

They will not have been moving quickly (and there I go, guessing already, but this matter is too pressing to leave unexamined, so we do what we can). They were searching. They did not know what they were searching for, but they would know it when they found it.

They ran their noses over the ground. They tested the air. They were not finicky—they never were—but found nothing right away as they entered the clearing. There will have been a jumble of smells inside the campground, a wild and tempting jumble. It did not matter how clean the campsites were; the four will have been able to pick up even traces of odor, plenty of it coming from the kind of high-calorie, sugary, fattening food that is rare in their world but common among humans. There followed a moment in which a game of chance was played. It was reminiscent of the famous short story by Shirley Jackson in which the town selects "winners" by lottery, and then, to ensure a good harvest, stones them to death. Here, the outcome may have been worse. With twenty-four occupied sites to pick from, they chose #16.

The four were bears, and they were the dangerous kind, *Ursus arctos horribilis*, the "dreadful bear," a name that preserves for us how the early explorers of the American West felt about it.

That there were four of them is perhaps the first feature of the evolving incident that was odd. Bears are loners; here in the latter part of summer, they ought to have been wandering, solitary and untethered. When they are together, it mostly has to do with reproduction, and that is what was happening here: This was a mother with three cubs, born the previous year. Considering the impact she made—for a few days, she was the most famous bear in the world—it feels as if we should know more about her than we do. Most of us have the idea, admittedly a vague idea, that bears live an easy life. It gives us the lyrics from the Disney film *Jungle Book*:

Look for the bare necessities, the simple bare necessities
Forget about your worries and your strife (yeah man!).
I mean the bare necessities
That's why a bear can rest at ease
With just the bare necessities of life.

The most dangerous kind of bear: a grizzly, with cubs. BRYANT AARDEMA/GETTY IMAGES

It is an appealing idea, but the female and her three cubs gave it the lie. She had led a hard life. Her weight was normal, but at the bottom of the range for "normal." In fact, by the general standards we apply to the species, she was, at 216 pounds, not large at all. The yearlings were similarly thin. The mother's mammary glands were producing a little milk, but only a little. Fat stores were marginal. She had both tapeworms and roundworms in her stomach and intestines, and her small intestine was inflamed from the parasite load. She had pneumoconiosis, which in humans would be called "miner's lung" or "black lung disease." It comes from inhaling dust, and she had probably gotten it digging a den, or digging food. Analysis of her hair revealed that she had lived almost exclusively on plants for two years; scientists who examined her concluded, "Ninety-two percent of the grizzly bears in the Yellowstone ecosystem consume a higher proportion of meat in their diet than did this grizzly bear." The immediate area had had a good crop of whitebark pine nuts, but she had not eaten any, or not many. When you belong to a species that has to put on loads of fat during the short period of warm weather before going into a long hibernation, it is a rough way to make a living. And she was still nursing.

So she had had a tough time. Given what followed, an idea circulated in the extended park community in the aftermath: that her behavior could be explained by stress. Being around the bear equivalent of three teenagers would not have helped. She had, in effect, just finally gone off the deep end. But, of course, we will never know.

As we have seen, the bears stopped first at campsite #16. There came, now, that instant in time that is the source of so much speculation, the moment when "a grizzly explores a person as something to eat." That line comes from the classic book *Bear Attacks: Their Causes and Avoidance*, by Canadian wildlife biologist Stephen Herrero. In it, Herrero brings all his vast experience to bear on the question of what makes these animals behave the way they do. "A psychology professor I knew," Herrero writes, "used to tell his classes, 'Studying nuclear physics is child's play compared to studying child's play.' What he meant was that the behavior of children—indeed, all animal behavior—is complex and very difficult to predict. Anyone who studies behavior soon learns this truth. It applies to bears as well as people." Since Lewis and Clark first encountered grizzly bears at the start of the nineteenth century, enough people have been attacked by bears that Herrero can draw some conclusions:

> *Grizzly bears usually enter camping areas at a walk, and at night. Before an attack, a person seldom sees any signs of aggression; such attacks, however, typically occur when there is too little light for a person to see. No single course of action can be recommended. If you are suddenly bitten while your body is bulging out of the side of a tent, then perhaps the normal startled yell will be enough to let the bear know that you are a person and not a salami. Most campground-marauding grizzlies still have enough "fear" of humans that they will flee when discovered.*
>
> *But if you are camping in a remote area with a small party, the bear may press the attack. Under such circumstances, playing dead would be akin to offering yourself to the bear. If you suspect that a grizzly is about to eat you, you must do everything possible to deter the bear momentarily so you can escape. All group members should shout at the bear. Throw things at or near it to try to distract it. Use every possible weapon or deterrent you might have.*

The victims, this night, tried more than one strategy. Only one of them worked.

The bears at campsite #16 confronted—or would have confronted, if the light had been better—what would almost count as a crowd in this part of the world. Here was the Fleming family: Maria Fleming in one tent, her father and sister in another. Also in Maria's tent were her boyfriend, twenty-one-year-old Ronald Singer, and their dog, described in the ranger's report as a "young puppy." The dog was not formidable, but Ronald was. The two, Ronald and Maria, were awakened when the tent, even though it contained two adults, a dog, and whatever they had with them, went sliding several feet across the ground. Three things happened then, more or less at once: Ronald felt something powerful and massive bite his leg through the tent fabric, Maria screamed, and Ronald started punching.

It was the one thing on this long night that went well. Herrero would certainly have approved. Ronald, an engineering student at the University of Colorado, did not talk to the media in the aftermath, but his mother spoke briefly to the *Denver Post*. His justifiably proud mother said, "He is strong! He was a wrestler in high school and it [his instincts] just kicked in. He started punching. He didn't know how many times he punched it."

Some of the blows landed, and the bear disappeared. Singer never got a look at the animal. He and the rest of their party left and drove to Cooke City, a mile west of the campground, where they managed to get through on 911. Ronald had his injuries treated at one of the hotels there, which makes sense: The lobby is lit, the door is unlocked, and, sleepy or not, someone is on duty. There is no hospital in Cooke City; there is nothing resembling even a clinic. He was able to ride to the hospital in Cody with the Flemings. In light of what happened later, news reports tended to describe his wounds as moderate, even minor, but they were not quite. On his lower leg were lacerations two and three inches long, with deep muscle damage. How deep? The bear's teeth dug divots out of the bone.

Now, bear attacks are not common to begin with. Even at this point in the event, we are already into the "vanishingly rare" range. As of late 2024, in its entire history, eight people have been killed by bears inside Yellowstone National Park, and the real number is actually seven; the very first attack, over a hundred years ago, was recorded in only one unreliable source, and we can disregard it with near certainty. The authorities would like everyone to be as cautious as possible in the park, but they will admit that the odds are strongly in your favor. On their Yellowstone "Bear Management" website, the National Park Service says it like this: "More people in the park have died from drowning (125 incidents) and burns after falling into hot springs (23 incidents) than have been killed by bears. To put it in perspective, the probability of being killed by a bear in the park (eight incidents) is only slightly higher than the probability of being killed by a falling tree (seven incidents), an avalanche (six incidents), or being killed by lightning (five incidents)." An alert reader will have noted, though, that the Soda Butte Campground is not in Yellowstone National Park; it is well outside the Northeast Entrance. Still, the numbers are not especially scary. According to the experts at the Interagency Grizzly Bear Study Team, in all of the Greater Yellowstone Ecosystem, bears have killed twenty people since 1892. The IGBST maintains a kind of scoreboard. Ominously, for several years now, they have been updating it every year, although as I write in late 2024, the bears are getting ready to hibernate, and we may make it through the year without a death.

It is generally assumed that it was the mother bear that launched the attack at Soda Butte Campground. She now did something bears rarely ever do under any circumstances. Getting pummeled by Ronald Singer was not enough for her. She moved to another campsite.

And there found another victim.

There, in campsite #11, was a tent occupied solely by Deb Freele, a fifty-eight-year-old retiree who had been fly fishing in the mountains. She was in one tent, her husband in another. "He snores, and he drinks and I don't like the smell of that," Freele later told *Slate* magazine. All was otherwise peaceful—until it was not.

"I had been sound asleep, and I had this sense that something was badly wrong, and it was bringing me out of my sleep," she remembered. "I was just becoming aware, and the bear clamped down on my arm. The tent was gone at that point. Then the bear bit down and held me there for a while. My back was to the bear and to my bear spray. The bear was driving me into the ground, and it was trying to pull me up."

She tried, first, and instinctively, to follow Herrero's advice: "The more I yelled, the more aggressive the bear got. I needed to let everyone know in earshot; I hoped people would get together and scare the bear off, but no one did anything. . . . The bear continued to bite. The more I yelled, the harder it got, and it still hadn't brought its paws into play. I was trying to think about what might happen next, and the thought of it was just horrifying."

So she went to the next, and standard, piece of bear advice: She tried playing dead. "I just went limp. Like a rag doll, didn't move a muscle, didn't move an eyelid. You can disassociate yourself from what's going on. I was only hoping I could get my bear spray. The other option was, how could I end it quickly?" If she could only get the bear spray. (People new to the Greater Yellowstone area may wonder what that is. Bear spray is a simple but powerful weapon: 8.1 ounces of capsaicin, the active part of chili peppers, which sounds tasty and harmlessly domestic, but is unbelievably irritating.)

If only she could get to it . . .

Playing dead did not work, either. The bear only bit harder, and harder, and Freele was reminded of what she had heard about lions in Africa, how they keep biting harder until the bone breaks. The pressure was like a vice, and she ultimately heard and felt a snap. In the surreal, hallucinatory weirdness of the whole experience, she heard sounds that she could not place at first. It made her think of the forest at her home in the Canadian province of Ontario, of the sound of worms moving under the leaves on the forest floor. Much later, she came to understand that it was the three yearling bears. They were not innocent bystanders in all this. They were jumping around like puppies whose mother had caught something interesting, and one of them now bit her on the leg. Freele thought she was part of a schoolroom lesson for the young bears: The mother was holding the upper half of her body "so she could teach her cubs to go for meat." She drew a conclusion with which Professor Herrero would agree: "I could tell it wasn't normal bear-seeking-food behavior. I figured, I am definitely prey."

That night, the bears were looking for a person to eat.

In the end, nothing Freele did contributed to her escape, except that her screams alerted the people in the next campsite. They then made a run for it, and could be heard locking themselves in their vehicle. It was the *click* of the lock that distracted the bear and caused her to move on. Again, news reports tended to describe Freele's wounds as modest, but as with Ronald Singer, they were not. She was taken first to a hotel in Cooke City, then to the hospital in Cody, Wyoming. The doctor who treated her said that her left arm had suffered a cut six inches long with deep muscle damage; that arm was broken and covered with other lacerations. She had surgery on the arm, was in the hospital for four days, and spent a year regaining use of one hand.

Of course she did. She had been attacked by a grizzly bear.

One thing Yellowstone locals noted, starting the next morning and continuing to the present day, was the response of everyone else in the campground, which one observer referred to as "an 'every man for himself' reaction." One element that no one has adequately considered, that I know of, is alcohol. As we have noted, drive-in campgrounds are beset, every night during the summer, by a certain percentage of people who go there to not just drink, exactly, but more like "get as drunk as possible." If the people in that campground were partying the way campers regularly do, not hearing a shout, or a scream, or a car horn would just be normal. Deb Freele's husband, after all, slept through the whole attack, and he was in a tent in the same campsite. The elevation would have worked with the alcohol to make lowlanders all the more drunk. Elevation has that effect: A single drink can render a light drinker inebriated, and the campground is over 7,500 feet above sea level.

There is also what we might think of as a "Pearl Harbor effect": That morning in Hawaii, when the Imperial Japanese Navy attacked, the response of nearly everyone was comically inadequate. They just could not believe it was happening. Something like that will have happened here, although some did try to raise a warning. When the people in the campsite next to Deb Freele locked themselves in their vehicle—the door locks inadvertently saving Deb—they drove around trying to warn people by honking the horn. They had children with them, so they were not going to open the doors again. They were ignored. As they drove, they missed the campground's west loop. Afterward, they regretted that—but amid the confusion and the dark and the fear, did they even know the loop was there?

So far, the same four bears—though really, just the one—had racked up two human victims. That number, two, is outlandishly high for bears. They are normally not interested in killing people. This one was. And she was not finished.

She and the yearlings walked through most of the campground and down that west loop. The other campers in the almost-full campground did not know it, but death had walked by, given them a sniff, considered them, and walked on, the youngsters following. Do they know that to this day? The bear stopped, at last, at the site occupied by Kevin Kammer, a visitor from Grand Rapids, Michigan.

Montana newsman Scott McMillion covered the event. "Kammer's site, number 26," McMillion wrote, in *Montana Outdoors* magazine, "was one of the most isolated. The closest campsite was 60 yards away. It was very dark out, and the nearby stream masked noises. Nobody saw or heard a thing." The investigators felt especially sympathetic when they learned more about him. McMillion continued:

> *Kammer, 48, had taken a break from a career as a medical technologist to be a stay-at-home dad to his four kids, the youngest just 9 and the oldest 19. Two of them attended a Christian school. His family chose not to comment for this story, but news accounts and comments from friends provide at least a partial picture. Dedicated to his family, he was the kind of dad who showed up at school board meetings when a decision affected his kids. When coworkers had a bad day, they could count on the affable and friendly Kammer to elicit a laugh. He liked fishing and camping and kayaking, fixing up the house, and relaxing in the hot tub.*
>
> *His camp was clean. His food was properly stored. He wasn't in the wilderness; he was in a campground, a few feet from his car, a place with toilets and picnic tables and improved roads. Though surrounded by wild country, this place was built for people. He did nothing wrong.*

But that is the thing about bears: You can never be sure what they are going to do. As we have noted, as much as anything else, what scares us about them is their unpredictability.

At about 4:20 in the morning, a Park County sheriff's deputy drove through the Soda Butte Campground and at long last brought everyone's vacation from reality to an end. With a loudspeaker and a spotlight, he woke them up and got them moving, out and away. When he got to Kammer's site, he found that the bear had done here just what she had tried to do with both Ronald Singer and Deb Freele. With Singer, she had met her match; with Freele, she had been distracted. Here, she had finished the job.

She reached in and attacked Kammer while he was asleep inside the tent, pulling him out by his head and shoulders through a hole she had ripped in the insect screen. She pulled him to a spot four feet from the door of the tent. Here, he bled to death; the pool of blood soaking into the earth at this spot told the investigators that. She later pulled the body ten yards away, and because she had so much time with it, much of the torso was gone. She and her offspring had eaten it.

They are beautiful creatures, but yes, there is a side to them that is all wrath.

Even the "safer" species of bear has it. The American black bear is, again, smaller and much less aggressive than the grizzly. In traditional media, in film and TV, black bears regularly appear as figures of comic relief: bumbling, playful, charmingly inept, filled with harmless mischief. Given his size and reliance on trickery over force, it appears that the iconic Yogi the Bear of Jellystone Park is a black bear, although it is hard to know what to make of Boo-Boo, who appears to have suffered some kind of growth hormone disorder. Yogi made his debut on the animated series *The Huckleberry Hound Show* in 1958, and got his own show in 1961, both, perhaps appropriately, sponsored by Kellogg's. Food defined him. During that era, a major pastime for visitors to Yellowstone National Park was to feed the bears. Black bears parked themselves along the road, and cars lined up with bags of marshmallows and the like. Feeding had been illegal in the park from an early date, but there was always a faction inside the Park Service that believed the public would demand it. There was also always a faction that hated it and thought it a travesty, and when it became possible to eliminate it, during the "ecology" fad of the early 1970s, they did so. Nevertheless, the park archives are filled with photographs running well back into the nineteenth century of tourists who stopped along the road and made black bears do ridiculous things to earn a bit of junk food. Out of this cultural dynamic, Yogi was born.

So, too, a charmingly silly Disney movie of this time, *Yellowstone Cubs*, which debuted in 1963. In it, the black bear cubs Tuffy and Tubby are separated from their mother, and hilarity naturally ensues. The film, clips from which are available here and there on the internet, saves for

An American black bear. The two bear species can be difficult to tell apart, but the black bear is often genuinely black and has a distinctly blond snout. JAMES HAGER/GETTY IMAGES

us scenes from the 1963 park, including a view of a more or less normal bear jam, complete with feeding antics. "Most of the first-time tourists just want to be friendly, but even the bears take a dim view of things when familiarity goes too far," explains the narrator, the cowboy actor Rex Allen, in a characteristic 1963 Disney, folksy voiceover where "view of things" becomes "view of thangs."

When popular culture gets hold of black bears, it turns them unthreatening. Even when a black bear is formidable, the culture turns it a little silly, as with California's celebrity black bear that got into the habit of breaking into houses around Lake Tahoe and eating just about anything it could get its mouth around. As often happens, multiple animals were actually involved in this lengthy string of incidents, and when the California authorities captured the main villain in 2023, it turned out to be a female with cubs. The public, guessing an aggressive bear would be male, had named it Hank the Tank during its long rampage. He/she turned into everyone's favorite; one of my students called Hank "a great American hero." Hank was a real-life Yogi, just more successful, and enormous. She got to be five hundred pounds, and in some photographs is shaped like a bowling ball.

Other celebrity bears have specifically been black bears, like Gentle Ben and Smokey the Bear. The species has more recently loaned its common name to a successful chain of coffee shops, the Black Bear Diner, where you eat surrounded by stylized bears. "The unique and clever bear carvings," the company says of its locations, "reflect the local flavor of each diner. You'll find bears as skiers, fighter pilots, grape-stompers and more."

But the real thing can be terrifying.

The black bear is, as noted, the one you are most likely to encounter during a visit to Greater Yellowstone. They are easiest to see in the area around Tower Junction and the Northeast Entrance Road. If this part of the park strikes you as . . . *different* from the stretches down by, say, Old Faithful, your impression is correct. Locals call this area the "northern range." The elevations are lower, and the ecology is altogether different; the result is long open meadows and sagebrush flats miles across. Because of the general lack of dense forest and the wide-open sight lines, it is possible to spot animals from far away. That black spot in the distance may very well be a bear. Furthermore, the Roosevelt area, every year, hosts at least one female black bear with one or more cubs, who parks herself in the area and displays the cubs as if posing for photos. There is one every year, a different bear usually, and sometimes more than one in a single year. The locals call her, or them, "Rosie," and the cubs are cute enough to melt the photographer's lens; when I first spotted Rosie's cub in 2024, for an instant I thought—I actually thought—that someone had lost a stuffed animal. Just look for the bear jam, the contemporary equivalent to the bear jam in that Disney film.

It may feel like you are in a zoo, but you are not. Bears are bears. They are large, and they lead large and dramatic lives. Seeing one under any circumstances is a thrill. And seeing one on foot in the backcountry, away from roads and civilization and any hope of immediate help, is the experience I keep talking about. Here is where you escape the fake life and meet the real. Just follow the National Park Service's rules for safety around bears: At the very least, do not get closer than one hundred yards to a bear of either species, or a wolf (for our numerous visitors from overseas, one hundred meters is about the same). Whenever you are on foot away from roads or villages, carry bear spray. The park concessionaires rent canisters, and they are on sale in lots of places in Greater Yellowstone. Here, also, is a piece of insider information that it is handy to have: The Costco in Bozeman normally sells a package of two for the exact same price as one in Gardiner, and sometimes even less. The National Park Service has more elaborate instructions on bear safety in the backcountry; it is, of course, on their website.

Black bears can and do kill with some regularity, although it might or might not have ever happened in Yellowstone National Park, keeping in mind that the landscape of the park is maybe twelve thousand years older than the mere "Park" designation, running back to the retreat of the glacial ice, and there were people and bears in it for most of that time. Black bears furthermore have a terrifying habit of targeting children.

Stephen Herrero, in *Bear Attacks*, has an entire chapter on this nightmarish topic, and it makes for some of the most awful reading in the book. As we have seen, grizzlies attack when they feel threatened, which apparently leads them, when they do attack, to go after full-size people and not the much less threatening man-cubs (although as with so many assertions about the habits of bears, saying so requires that we read their minds). What appears to happen with black bears is that local food sources, like berry crops, may occasionally fail, and when hunger makes them desperate enough, they begin to take an interest in humans as a source of calories. They then attack those humans who represent the easiest prey.

Hence, the attack on little Carol Ann Pomranky in 1948. She was three, and her family was living in a house in what was then the Marquette National Forest, near Sault Sainte Marie along the south shore of Lake Superior. Disaster struck from out of the summer sky, as Carol Ann's mother told an investigating ranger. The day, July 7, was hot, and

Carol Ann was playing outside. "My first indication that anything was wrong was when I heard Carol Ann utter a cry of alarm," she said.

> *I ran to the back door, which was open with the screen door closed.*
>
> *Carol Ann, when I reached the door, was on her hands and knees on the porch with one hand touching the screen in the door, and a bear was about midpoint of the three steps of the back porch. The bear growled, showing his teeth, and grabbed Carol Ann with his mouth and pulled her off the porch onto the ground where he picked her up by the arm with his mouth . . .*
>
> *I ran for my husband's pistol, a .32-20 Colt revolver, but in my excitement could not load the pistol. I ran to the door, but the bear and child had disappeared into the bushes.*

A search party was organized, and a dog led them to Carol Ann's body, partly consumed. The bear was killed a short time later. It was, of course, a black bear; the grizzly is a western species, and never lived in Michigan. The bear was physically normal, but thin, and the berry season was some weeks away. The best guess is that it was driven to this behavior by hunger.

Herrero's recommendations are different for black bears, compared to grizzlies: If you find yourself under attack by one, he believes that you should not play dead, as with grizzlies, but should do everything in your power to resist, using any weapon available, including your fists. He also tells the story of an attack by a black bear on an unnamed ten-year-old girl in British Columbia in 1976. The report of the incident, in the form of a memorandum by the British Columbia Fish and Wildlife agency, calls the location a "lonely hut" in the woods—almost like a fairy tale, but entirely real. A bear approached her as she was fetching water from a creek. She picked up an axe, hit the bear twice, and ran for home. The bear followed, knocking her down and breaking ribs. Once home, the bear tried to follow her inside. She hit the animal with the 1976 equivalent of bear spray: She threw a pot of boiling water in its face. That was enough. It ran. Herrero is filled with praise for her actions (as is pretty much everyone else who has heard the story since then). He thinks the bear was trying to eat her. Under those circumstances, the thing to do is fight back.

The most recent edition of Herrero's book, from 2018, continues to call black bear attacks rare, as it did when the book was first published in 1985, but . . . I don't know. Wikipedia has a page titled "List of fatal bear attacks in North America." For all the assertions people make about how very uncommon bear attacks are, the page gets updated with some regularity. It includes, just counting the fatal attacks since Herrero's first edition, eight deaths by black bear in the 1990s, sixteen in the 2000s, eleven in the 2010s, and eight so far in the 2020s. These include a seventy-one-year-old woman who was killed by a black bear in 2023 in the Sierra Nevada town of Downieville. She became California's first-ever known victim of a fatal black bear attack.

So, Smokey the Bear is not really your friend, either.

When encountering a bear in the backcountry, away from civilization and whatever aid that might give, it is best to be cautious, without allowing yourself to be terrorized. Assume any bear is a grizzly, especially since they can be difficult to tell apart, and many believe one hundred yards is not enough distance to keep from a grizzly, especially a mother with cubs; they go to extraordinary lengths to protect their young. The bears have distinct profiles at the head. Grizzlies have a concave facial profile, while a black bear's profile is long and straight, closely resembling that of a bull terrier. A grizzly's ears are smaller, more round, and disconcertingly cute. Most strikingly, a grizzly has a distinct shoulder hump and claws so large they are visible from a surprising distance. These adaptations are not for ripping you up; they are for digging for food, one of a grizzly's favorite activities. They also spend a surprising amount of time just grazing.

Color is not the best way to tell them apart, but the difference can be striking. The grizzly gets that name from its grizzled fur, which is streaked with gray up and down its sides. When a grizzly runs—and they can run faster than most dogs—the flesh underneath that fur rolls like a wave, and it may remind you of something hard to place. It takes quite some time to pin down the comparison, but it comes at last: The shimmering ripple is like the aurora borealis—the northern lights. If, on a long trip through the West, you were to leave Yellowstone and drive north, you would, soon enough, reach places where the aurora is a more regular companion than it is in the United States. You can see it in the Canadian province of Alberta; it appears in the Rockies—where, appropriately, grizzlies are as at home as they are in Yellowstone—or lighting up the sky along the northern rim of the prairie. It can shine so brightly that on summer nights when the sun sets late and rises early, darkness never really falls. When that happens, the aurora may present itself as a wall of light that looks like a curtain, a glowing curtain across the northern horizon, and as you watch, streaks of brighter light flow across that curtain, from one end to the other, just a bit faster than they would with the curtains in your house if you set them on fire. And, yes, that is what the side of the grizzly looks like when it is running: fire in the heavens.

Best give it room. What is rippling there is the outer few layers of a body that, in Yellowstone, may weigh 700 pounds. In Alaska, where they get more food, they can reach 1,200 pounds or more.

For at least a century, visitors to the western United States, going to places where bears are likely to turn up, would hear advice and opinions about them that today is quaint at best. Now that the advice is old enough that it is no longer likely to get people killed, it can even be cute. Visiting a national park with grizzlies in it at any point up until roughly the late 1960s, this was what you would hear: Confronted by people, they are natural cowards. They are terrified of fire. Their sense of smell is acute, but their hearing is weak, and their eyesight is comically bad. Their favorite foods are basically anything that is horrifyingly rotten, although they are easily satisfied and never picky. They do like honey more than anything in the world.

What we get from the traditional lore is an animal that seemed to owe a good deal to old cartoons: Baloo, Winnie the Pooh, and Mr. Magoo.

Over the course of roughly the last fifty years, we have learned much. We know more now than we did then about grizzlies and their aggressiveness toward humans: why they attack, how attacks tend to go, and what can be done to avoid them. Sadly, we know these things partly because there have been attacks, quite a few of them, although the number is not increasing drastically, as some people believe. Glacier National Park, on the Canadian border to the north of Yellowstone, has hosted grizzly bears for all of its existence, and it is true that there were no attacks in that park for that entire existence, until a single night in 1967. Before that night, the bogus "wisdom" above was readily available there. Nearly everyone, it seemed, believed grizzlies were natural cowards that would always turn tail.

That night in 1967 can stand as a rough starting point for the education we have had, over the last half century, in bear attacks. In many attacks, as we have learned, the bear is not to blame. Often, the bear is "habituated." Two of Yellowstone's bear experts, Kerry Gunther and Travis Wyman, explained habituation in a 2008 issue of the in-house journal *Yellowstone Science*, published by the National Park Service. "Human food conditioning is defined as the attraction to human foods or garbage due to prior food rewards giving positive reinforcement. . . . Human food conditioned bears are almost universally considered a problem and dangerous to personal property and human safety by most bear management agencies. Most bears conditioned to human food eventually become aggressive in their efforts to obtain human foods and damage property or injure people in the process. Then they must be destroyed by managers." If you ever had a class in psychology that reviewed the work of Ivan Pavlov with his famous conditioned dogs, and BF Skinner's "operant conditioning," you will be familiar with the basic ideas. Professor Herrero, who may be the reigning expert in the world on the subject of habituation and bears, explains it like this:

> *A new noise, sight, or smell usually catches an animal's attention, and it responds in some way. But if the stimulus is presented repeatedly and nothing good or bad happens, then the animal becomes used to the stimulus, and attention and response wane. The animal has become habituated to the stimulus. If a bear regularly encounters quite a few people, but doesn't get food from them and isn't harmed, it simply gets used to people, will tolerate them at closer distances than before, and will sometimes ignore them. Such a bear is habituated to people.*

> *A habituated bear that also eats people's food or garbage behaves differently from a bear that is only habituated. Such a bear forms a simple association—"people" may be followed by "food." . . . While habituation may occur without food-conditioning, the food-conditioned bear is almost always somewhat habituated to the smell or sight of people.*
>
> *Habituation combined with food-conditioning has been associated with a large number of injuries. Inside the national parks it is probably the most frequent circumstance associated with injury.*

The essence of the situation is that bears do not belong in campgrounds, or along roads, or inside the pantry of your expensive South Lake Tahoe VRBO. Put them in such places regularly, and a clock begins ticking. Think of it as the tyranny of statistics: Put a bear in contact with people, and the chances of a dangerous encounter, for whatever reason, go up considerably. The more regular the contact, the greater the chance; when the bear is in contact all the time, the chance spins upward toward certainty.

The rangers who run national parks no longer allow such habituation to develop. Over the course of, roughly, the second half of the twentieth century, parks with bears in them were elaborately redesigned. Incredibly, you used to be able to see bears feeding at garbage dumps inside Yellowstone National Park; there were even bleacher seats, and a ranger would deliver a lecture on bears while the grizzlies came in to feed on the day's garbage. As we have seen, the bears were fed beside the roads, too, and the rangers during this time mostly hated it. Paul Schullery, a longtime Yellowstone NPS ranger and former park historian, wrote a memoir about working in the park in which he shares the ranger's point of view:

> *What I've seen in Yellowstone has convinced me that feeding wild bears, in dumps or along roads, is a stupid, ugly, typically human thing to do. What bothers me most is not so much the people who get hurt but what it does to the bears. Hundreds of people were clawed or scratched in those days . . . but look what they were doing: ignoring all sorts of warnings; smearing jelly on a child's face so they could*

photograph the bear licking the child; placing children on the bear's back for a picture; feeding bears film wrappers, cigarette butts, ice cubes, cherry bombs, and even food; running over an occasional cub . . . in short, doing everything to test the forbearance of an incredibly patient providence. Providence frequently took the form of a mama black bear who finally had had too much and took a swat at the hundredth citizen of Poughkeepsie to make a grab at her cubs that day. Then the rangers would be called to destroy the "dangerous bear."

It was a travesty, and when history turned in their direction and they finally could do it, the authorities got rid of it all. The dump feeding shows were shut down, the dumps themselves eventually closed, and the beggar bears chased away from the roads. Today, when you see a bear in Yellowstone, you are almost always seeing a truly wild animal. Thank the garbage men, among others. Inside the national park—and regularly outside the park, too—trash is not treated with the nonchalance normal to human places. It is instead treated like jewelry. The garbage cans are designed to be impregnable to bears, even to the largest of them—in other words, impregnable to animals that can tear the door off a car with ease.

Both kinds of habituation—to food and to people—were common in Glacier National Park in the 1960s, where situations had developed that rangers and bear biologists would today regard as insane. But that kind of thing was common wherever you found bears. There were just fewer bears then, so the madness was less widespread than it might have been. It was here in Glacier that history turned, and the relationship between bears and people changed. Yellowstone and Glacier would both change as a result, and in the end very much for the better, although it took decades for the relatively happy ending to unfold.

During the summer of 1967, one bear in particular made a great nuisance of itself in the Trout Lake area in Glacier National Park. It had appeared earlier that summer at Kelly's Camp on the north end of Lake McDonald, on the other side of a high ridgeline from Trout Lake. It ("it" later turned out to be "she") was a sick-looking, abnormal bear, with thin hair that was missing in patches along the backbone and an odd misshapen head. Reading descriptions and the reactions people had made the bear sound like the banjo-playing kid from the movie *Deliverance*. Her behavior toward people was menacing from the start. She fed regularly on garbage behind the camp, and found any motion inside the buildings annoying; she would actually attack the building, on one occasion coming close to tearing a door down and attacking the inhabitants. She visited the camp fifteen times, terrorizing people on nearly every occasion. When she moved over the high ridge to Trout Lake, she continued with the same kind of behavior. She developed a habit of running people out of their campsite (the same campsite, repeatedly), chasing them away or up trees, then tearing the camp to pieces and eating all the food, or as much as was left after her fits of violence. She did it to a pair of teenage boys; she did it to a honeymooning couple; she did it to a whole troop of Girl Scouts.

According to Jack Olsen, who wrote the bestseller *Night of the Grizzlies* about these events, the bear had another odd habit: "There were times when the peculiar animal would follow campers for hundreds of yards, always staying 20 or 25 feet away, and scare them half to death. Almost always the victims of such encounters berated themselves later, the tenderfeet for not knowing that grizzlies are relatively harmless and the oldtimers for realizing it and still being afraid." An awful lot of people really seem to have believed that. However, there "was something about this persistent grizzly that alarmed even the most knowledgeable. Grizzlies had been snooping in and out of the campsites of North America ever since the first primitive man had pitched the first camp, but they had not made their intrusions while the campsites were occupied and certainly not while people were in the middle of meals and other activities. The oddly shaped grizzly did not seem to know fear. It stormed into camps and bowled over fire tripods, tents, and packs. It stayed exactly as long as it wanted to stay. It ignored the shouts and screams, and sometimes the rocks, of annoyed and displaced campers."

Even during that era, the government rules were clear: "Such a bear must be shot, but somehow the skinny animal managed to remain alive." The bears were like birds and squirrels at picnic tables. They had never killed anyone in the history of the park, so the misbehavior was tolerated.

The government rules were also clear on how concessionaires were supposed to handle their garbage. Thirteen miles from Lake McDonald is The Loop, a sharp turn in the Going-to-the-Sun Road; from here, four miles of hiking takes you to Granite Park Chalet (which is thus well into the backcountry, unlike any of Yellowstone National Park's hotels, even the most primitive). It has always been a strange place. Built by the Great Northern Railway in the early years of the twentieth century, it looks, from a distance, like a scene from *The Sound of Music*. Close up, one sees how primitive it is. During the "ecology" fad of the 1960s, it was wildly popular, especially when people learned that a hiker could nearly always see grizzlies there. Sadly, there was a reason.

Garbage, as you might have guessed, was the reason. The authorities had purchased an incinerator for the chalet the year before, for the princely sum of $84. It was only "as big as four or five shoe boxes," and could not begin to handle the constant flow of garbage. The chalet was trapped in a relentless feedback loop: The more popular it became—so many people were coming that overflow guests had to sleep on the floor—the more garbage it produced. There was no other place to put it, so the managers dumped the garbage where it had always gone, in a nearby gully, which lured in more bears, which made the place more popular, which led to still more garbage. Three years earlier, the government had established a primitive campground a few hundred yards downhill from the chalet. It was there to handle the overflow of people who hiked in, then found that there was no room for them at the inn. Summer 1967 at the chalet fell into a regular pattern. Every evening, the same bears showed up on the same trail, the one heading from the campground to the chalet. It was their regular commute, leading them to the gully-dump. The garbage was drawing grizzly bears right by the campground. It was all just kind of tolerated, by everyone, because no one had ever been hurt.

Bear managers in various agencies will read descriptions of what happened that summer with amazement, and in fact anyone who knows the basic facts about what happens with habituated bears will look with wonder on an era when such audacious lunacy was allowed. But look: We still do it. All over the park system, there are animals that are habituated to people, and food-habituated, too. The foxes at Channel Islands

National Park are infamous for begging, as are the coyotes in Death Valley, and there is a state park in South Dakota, Custer State Park, where people go just to see the burros that stick their heads through your car window, demanding handouts. In Yellowstone, if you climb Bunsen Peak, you will find, at the top, a ruthless gang of chipmunks that exploit their adorability as a means of extorting treats (check out the summit logbook, where you will find portraits of the animals drawn by visitors with, and without, artistic talent). At the picnic tables in various places around Mammoth—this can happen anywhere, but it seems to happen at Mammoth especially—ravens will walk right up to you, making those weird vocalizations you expect, given their vocabulary (*blorkel* is one, approximately; *glorck* is another). The ravens are not exploiting their adorability at all, but the goal is the same. For years, marmots would emerge from underneath the boardwalks at Old Faithful and beg; they even lost all their hair at one point, it was thought because of a diet too heavy in Twinkies. And at picnic areas all over the park, there are ground squirrels and songbirds, perhaps thousands of them, that make a living off visitors, carrying on Yogi's tradition.

A Bunsen Peak chipmunk, in the summit logbook.

A habituated raven. This one spent every day outside the Tower Junction restroom.

The rangers do what they can to stop this activity, but they are busy and have to prioritize. Feeding animals is illegal and nearly always harmful to the animal. There are signs everywhere reminding people of that. The animals, however, have us under their control. That is an important thing to consider, and it applies to beggar bears, too. We are not in charge in that situation. The animals, at some level, realize that they have this terrific power over us. The funny-looking multicolored creatures that get out of the rolling multicolored boxes, they know, have food, and if they strike the right pose, they can use it as leverage. Even the most enlightened visitors are unable to help themselves when a chipmunk jumps up on a boot and poses as if for a photograph. The chipmunk gets the Cheeto it came for. If one of these animals were ever to seriously injure someone, they would go to the top of the rangers' list of priorities, but these are animals weighing, often, a few ounces. Nor is feeding them instantly destructive. It is not the worst thing people do to animals, but it is an especially bad idea when the animals are bears, and dreadfully bad when the bears are grizzlies.

The great change began on the night of August 12–13, 1967.

In both cases, the park's concession employees were involved. Long ago, I was one myself. The businesses in the park are run by a small number of large corporations that are "conceded" the right to do business in exchange for certain considerations (the government gets a share, for one). These concession companies need to find an instant staff every year to fill positions that are very much not career-track jobs. They have traditionally relied on three groups: retirees, college students, and bum-luck drifters. Be patient with the person in the ill-fitting uniform trying to help you; that person may have arrived in the park even more recently than you, and may not speak much English (recruiters fly all over the world to fill these positions). The law-enforcement rangers find the students to be a handful, and always have; inexperience and exuberance lead them into the most absurd misadventures. Romance blossoms pretty regularly, and was involved in the events of that night (it was not to blame). Nearly everyone comes through it unharmed, and for the rest of their lives, they look back on that summer, or string of summers, as the most improbably perfect of their lives.

That is the kind of summer probably everyone was having in the group of five friends who climbed the high ridge and descended to Trout Lake on the night of August 12–13. They were in a hurry, a normal feature of that life; the late sunset will have helped them find their way, "late" for most of them because Glacier is on the Canadian border and so is much farther north than, for instance, San Diego, where nineteen-year-old Michele Koons was from. She was part of that group of five friends, all concession employees and most of them college students, who hustled over the ridge with, we can guess, the kind of frantic energy characteristic of such outings. The climb is two thousand feet over two and a half miles, and so can be challenging, particularly given the elevation. These five were untroubled, and even were carrying, literally, a dog, a mongrel puppy abandoned in the park and adopted by one of the party. When he tired, they picked him up and cradled him like a baby. Today, it is illegal to take a dog camping like this, in both Glacier and Yellowstone; dogs will see a bear and want to fight it, and so they are not allowed in the backcountry, for this and other reasons. The dog, however, played no role in what followed.

Michele shared the burden. She worked in the gift shop at the Lake McDonald Lodge, and so had been to Trout Lake more than once. She knew there were bears in the area regularly. She was, her boss later remembered, "a blessing, a girl with a zest for life." She was having, it appears, a typical park employee's summer, the kind you remember forever, the kind that makes the rest of life feel a little disappointing.

Their campsite was the very one frequented by the strange-looking bear. They were not worried, and might or might not have known anything about the possibly diseased and definitely weird grizzly that had made this her literal stomping ground since leaving Lake McDonald. They arrived at the site with time left to fish, but Michele volunteered to do the work of making camp, cooking hot dogs, and watching the puppy while the others caught fish to round out dinner. It was a happy scene . . . until the moment when, over a dinner of densely odorous fish, Michele looked up and screamed.

There she was, the strange, sickly, half-deranged grizzly, a shadow emerging from the trees not ten feet away.

They ran. In a chaotic pell-mell jumble, they ran down the lakeshore. Someone thought to gather up the dog, which was good, because he would have wanted to either fight or play, and he would have been killed. The bear did not follow. She went to work eating their dinner and, rather pointlessly, demolishing one of their packs.

When the bear at last departed, they debated. Hiking out in the dark looked like a bad idea, since they had only a single, small flashlight and the bear had left in the direction of the trail that represented their quickest route out. They decided to stay; after all, no one had ever been killed by a bear in Glacier. They made a new camp at the spot where they had regrouped after fleeing, arranging sleeping bags in a semicircle around a fire that they kept roaring as a theoretical deterrent to the bear. At length, the five friends drifted off to sleep: Paul Dunn, Denise Huckle, the brothers Ray and Ron Noseck, and Michele. The eldest, by the way, was twenty-three, the youngest just sixteen. The dog, Squirt, belonged to Denise, and the two huddled together now for warmth and security.

For what happened next, we can turn to the description by Jack Olsen, whose *Night of the Grizzlies* made this one of the most famous bear encounters in history:

> It was 4:30 and the fire had fallen to embers again when Denise heard a splash and narrowed her eyes to peer into the night and saw

a bear coming at a lope straight from the shoreline toward the center of the camp. When the bear was four or five feet away and she could make out its head and upper body clearly, Denise pulled the sleeping bag over herself and Squirt just as the dog began a high-pitched squeal. Lying perfectly still inside the warm bag, the terrified girl heard a ripping noise that sounded like shredding canvas, but then there was a silence broken only by the deep breathing and grunting of the grizzly.

Paul Dunn woke up to see the huge form of the bear standing next to him. Noiselessly, the boy slithered back into his bag and tried to remain absolutely still. He heard the bear making more sniffing sounds, and suddenly he realized that the sniffs were getting closer. Then something crunched into the bag and took a firm grip on his sweatshirt. Instinctively the boy threw back the flap and scrambled to his feet, slamming into the bear in the process.

It might as well be a horror film. The bear reared up as if to attack. Dunn ran to a tree and climbed, cutting himself all over in his mad rush. When he got past thirty feet, he looked down to see the bear circling like a shark. It did not follow him up the tree. Grizzlies normally do not.

Denise Huckle and Ron Noseck were in sleeping bags side by side. Dunn's flight distracted the bear long enough for them to take action:

Noseck told Denise to run, and when she did not move he yanked her from the sleeping bag and gave her a shove. The couple ran about 50 yards in the direction of the original camp, and as they ran they heard Paul Dunn shouting down from his treetop. They reached a slight incline and stopped, gasping for breath, as the puppy came bounding up. Ron boosted the girl up a tree, threw the dog to her, then shinnied up a tree of his own.

From his observation point almost directly above the camp, Paul Dunn saw everything that happened within the small circle of reddish light thrown off by the dying fire. He saw Ron and Denise run down the shoreline, followed by the puppy, and then he saw the grizzly walk toward Ray Noseck's sleeping bag and begin sniffing rapidly. When the bear turned momentarily toward Michele Koons' bag, Ray came out of his own as though shot from a gun and headed down the lake toward Denise and Ron, shouting as he ran, "Get out of your bag and run for it!"

Paul hollered at Michele. "Get out! Get out! Unzip and get out!"

The bear clamped its jaws on the side of the sleeping bag and Paul heard the girl scream. When the animal raked the bag with its claws, Michele cried out, "He's ripping my arm!"

"Michele!" Paul shouted. "Get out of your bag! Run and climb a tree!"

"I can't," the girl screamed. "He's got the zipper!"

Then the defenseless girl shouted, "He's got my arm . . . my arm is gone! Oh, my God, I'm dead!"

Paul Dunn saw the bear lift the sleeping bag in its mouth and drag it out of the circle of fire and up the hillside into the darkness. He shouted down the lake to the other three, "He's pulling her up the hill!" and then, "She's dead! She's dead!"

We can only guess what happened to Michele. The four others stayed in the trees until dawn, when they climbed down and, partly running, partly stumbling, they climbed over the ridge and went to find help. A day later, rangers shot an old female grizzly that matched the description of the bear that had been terrorizing the campground all summer, and it had human hair in its stomach. Ironically, they had actually intended to earlier, but it was a dry, hot summer, and the rangers were all fighting fires. Michele herself was found a hundred feet from the site of the attack. The bear had come to associate people with food, and now she took that association to its limit—quite a bit of Michele's body had been devoured.

But that was later. This night was not over yet.

At Granite Park Chalet, earlier that evening, two other park concession employees had arrived, not long before sunset. They were Roy Ducat, just eighteen, and Julie Helgeson, the elder of the couple at nineteen. Roy was a sophomore already at Bowling Green State, and had worked as a lifeguard; as the college-student concession employees normally were, he was in wonderful shape. In Olsen's description, Julie was every bit as appealing as Michele Koons: She "was a lovely slender girl with brown hair and blue eyes and a deep interest in nature. At 19 she was two years out of high school, where she had been a pompon girl, a singer in the school choir and a class leader. Now a sophomore at the University of Minnesota, she kept up her active life in the church. Her father liked to describe her in a short phrase: 'a beautiful, bubbling girl.'"

Today, the concession company she worked for would find a way to put this paragon in front of the general public, as a maître d' or a waiter—the park aristocracy, because of the tip income. But Julie had somehow gotten stuck in the bowels of the laundry. She may or may not have cared. She had been in the park for two months now, and this was her first backcountry camping trip—but was this really "backcountry"? A first "overnighter" is a rite of passage in the world of the summer employee. They were certainly a long way from the road. The pair were doing it the classic summer-employee way: They had left late, picking up sack lunches from their employer's kitchen and hitchhiking to Logan Pass, where they headed up the seven-mile Highline Trail (hitchhiking was the traditional way to get around; you held up a sign that said "Park Employee: Harmless," and were generally picked up in a few minutes). They made camp at about 8:00 p.m.

Roy, we know, had heard about the bears that visited here every night to dine on the garbage, and so presumably had Julie. They had also been through the mandatory Park Service safety lectures at the start of the season. They were not worried. When they arrived, they spoke to a married couple named Janet and Robert Klein. The Kleins, along with a twenty-year-old man named Don Gullett, were camping next to a disused cabin that stood beside the trail between the chalet and the campground; this location put them a comfortable distance away from the bear traffic. They asked if Roy and Julie were concerned at all about the bears that walked past the campground where they were planning on spending the night. Roy and Julie's response? They laughed. "Oh, that's nothing to worry about," Roy said.

In the middle of the night, close to the same time that the bear—the sick bear, an entirely different bear—showed up at the Trout Lake camp, far to the southwest, Don Gullett woke up in his sleeping bag next to the cabin. He was still half-asleep when he made out, to his alarm, the dim outline of a man standing at the foot of his bag. But the man was not a threat, because he now fell to his knees, and then fell flat. His arm had been hanging limp, and on his pants leg, blood was spreading. It was Roy Ducat, and he was in shock. "A bear got ahold of me," he said in a rush, nearly babbling. "I tried playing dead, but it didn't help. He dragged her off. You have to go after her! Oh, please, forget about me! The bear dragged her away." They looked upward toward the chalet, perched above them on the mountainside, its windows discouragingly dark. Robert Klein tried to signal the people there with a flashlight, while his wife and Don Gullett did what they could to comfort Roy, who was bleeding severely enough that Janet Klein was convinced he would bleed to death. Eventually, a guest at the lodge noticed, and an ad hoc rescue party of thirteen started down the hill.

Jack Olsen later reconstructed what had happened a little earlier (at about 12:45 a.m.) to Roy Ducat and Julie Helgeson:

> *As the little group waited for the help that they now knew was on its way, Roy Ducat slipped in and out of panic but never out of consciousness, and he remembered clearly what had happened. He had been sound asleep when all at once he heard Julie Helgeson, lying nearby, calling to him to play dead. While he was still trying to figure out what she was talking about, a single blow from a huge paw knocked him five feet away from her. Roy had landed on his stomach, but he could see Julie out of the corner of his eye. Then he felt something bite deeply into his right shoulder, and with a tremendous exercise of his will, he neither cried out nor moved. The biting stopped, and Roy opened his eyes long enough to make out the shadow of a bear standing on all fours above the helpless girl and biting into her body. He shut his eyes tightly in time to feel the bear return, plant its feet firmly in the small of his back and begin chewing on his left arm and the backs of both his legs. Still he remained silent, and once again the bear lost interest and returned to the girl.*
>
> *Now Roy could hear bones crunching, and Julie screaming out, "It hurts!" and "Someone help us!" He realized that the girl's outcries were*

receding down the hill, and he thought the bear must be carrying her off. When the screaming stopped, he jumped to his feet and ran uphill as fast as he could, unaware of his own pain, and slumped alongside Don Gullett's sleeping bag. All Roy Ducat could think of was Julie Helgeson, and her helplessness and his own helplessness in the face of the attack.

The rescue party found them and carried Roy Ducat to the relative safety of the chalet. Frowsy though it was, a great many successful people and professionals visited Granite Park Chalet (it was really the park, of course, that drew them). Tonight, the guests included a surgeon, a nurse, and two other doctors. They worked on Roy. It was clear he would require surgery, but he would be okay.

What happened next, however, has been a source of annoyed commentary since the events occurred, over a half century ago. A rescue party was preparing to go find Julie. A ranger-naturalist who also happened to be staying at the chalet stopped them. She thought it was too dangerous. Julie might have been rescued, but for the delay. In fairness, she had a point—an extraordinarily dangerous bear was wandering around out there in the dark. And the decision should more appropriately have been made by the park superintendent or the chief of law enforcement. The naturalist was just out of college.

Two hours after the incident had begun, at 2:45 a.m., a helicopter arrived. It carried Roy back to civilization, an astoundingly quick rescue, under the circumstances. It also brought an older ranger, one armed with a .300 Winchester Magnum rifle, a step in the right direction, but it was still terribly dark out (the moon was at 57 percent illumination and waxing, better than nothing, but no match for sunlight). They marched nervously downhill. They found chaos at Roy and Julie's campsite, and a pool of blood. From it extended two trails of blood, one headed uphill, one down. They followed the trail downhill, lost it, then picked it up again. They found a purse covered with blood ("empty," according to Olsen, "except for a single dollar bill"). Finally, 120 yards from the campsite, they heard an eerie sound. They could barely make it out. At last,

"From off to the left and slightly further down the slope, they heard a muted cry for help."

Julie was lying face-down: "Her body was ripped and torn, and she was covered with blood. To the first observers, it appeared impossible that she could be alive. But when Dr. Lindan came running up and dropped to his knees over the prostrate form, the girl began to move her lips, and then everyone could hear her say, 'It hurts.'"

They carried her back to the chalet and laid her out on a table. Both her left and right chest were punctured, the doctors saw; she had what soldiers call a sucking chest wound. One lung was collapsed. The wound that caused it, or at least the worst of them, was an inch and a half long. There were others, and they were audibly hissing. The doctor's first move, in such cases, is to cover the wound, or wounds, enough to restore the seal, allowing the lung to draw in air, but the major wound was too large. The surgeon in fact thought that so much time had passed that they could not have saved her even if they had been in a hospital, and not a barn made out of rocks cemented together, stuck to the side of a mountain in deep wilderness.

The doctors did what they could, which was very little. Julie was sinking fast. Every bit of energy she had was going into her hopeless effort to breathe, and the air kept hissing away.

Among the unusually accomplished guests of the chalet was a Jesuit priest, who now approached and spoke quietly to Julie. She responded in a whisper. When it was clear what was happening, he asked the nurse for a glass of water. The nurse told him not to let Julie drink it, to instead give her a rag to suck on, but the priest did not want the water for that purpose. Instead, he baptized her, and as he said the Act of Contrition, she tried—to the amazement of all—to follow along. Then, at last, the tortured breathing stopped.

Chapter Two

Lessons Learned

The Bears II
It was, however, the beginning of a new era.

An absolutely nontrivial part of the matter was that both Julie and Michele were terribly young and strikingly beautiful. Surviving photographs of the women do not do them justice; only all of them together give some sense of how dazzling they both were.

And they were, both of them, very nearly the most blameless and tragic victims imaginable, but that is the way it often goes. The bear never seems to attack a villain.

However, in one way they did not die in vain. No one, in the future, could imagine that the grizzly bear is a harmless coward. From their deaths, we can date the beginning of modern bear management.

The changes did not, of course, just affect Glacier. Today, in Yellowstone, we shape our own behavior in many ways to accommodate the bears, and we also try to get them to accommodate us, to the extent they will do it.

At the time the attacks happened in Glacier, garbage was still disposed of in open landfills inside Yellowstone National Park, and as we have seen, the grizzlies fed on it. Dozens of them came down every evening to dine on the leavings of the park's human visitors. It is, believe it or not, still a matter of debate among wildlife biologists whether the practice of eating garbage was harmful to the bears. They can tolerate bacteria that would seriously inconvenience a human. Bears are loners, and packing them together like that makes it likelier they will fight and also spread disease to each other. At the same time, as the pioneering bear biologists Frank and John Craighead argued back then, it is perfectly natural for bears to assemble into squad-, platoon-, and even company-size forces at places like those waterfalls you likely have seen films of, where the bears gather to eat salmon. The aesthetics at the dumps, however, were appalling, so the Park Service closed them.

At about the same time, the beggar bears were run off from the roads. It was generally agreed that both garbage and human food were dangerous, for both bear and human. Today, you are required to store food and anything smelling of food, or like food, in a bear-resistant container or a vehicle. In the backcountry, packs or other containers with food in them have to be suspended; inside the park, poles are provided, about twenty feet off the ground. It is not just food that the camper or hiker needs to worry about, but anything that might serve as an "attractant." This somewhat loose and slippery category definitely includes dirty dishes and cookware, beverages, coolers, food containers (a pizza box is, from a bear's point of view, an odor-bomb, pure corrugated torment), grills, bug repellent, sunscreen, and stuff for the dog, including food and even water bowls. The vagueness of the boundaries around what is defined as an attractant has led to unfortunate flights of paranoia. For years, people were afraid that the odor generated by sexual intercourse would draw in bears. There were stories that owed a lot to slasher movies of the 1980s, in which the young lovers give in to temptation and are killed and eaten as punishment. They led to a great many whispered conversations in the tent, in the backcountry, in the dark, veering between dismissive and frantic. No one really knows if that one has anything to do with reality. Menstruation, too: For a couple of decades, between roughly the 1970s and the 1990s, it was common wisdom that a menstruating woman was running a great risk in the backcountry, and looking at backcountry advice websites as I have been writing this, I still see it regularly today. Toothpaste is another possible attractant, and a good deal more likely to be real, along with Fig Newtons. Not the most odorous of foods, the humble Fig Newton has still been known, in Yosemite, to motivate bears strongly enough that they will tear cars apart to get at them.

This is how we got the following old saying: "A pine needle fell in the forest. The eagle saw it. The deer heard it. The bear smelled it." That is supposed to be a Native American proverb. It might or might not be, but it does get close to the truth.

Needless to say, we do not feed bears on hotel garbage anymore. Indeed, we bend heaven and earth to avoid providing a grizzly with a meal, or really to intervene in their diet at all. When an animal dies near a trail, the trail may be closed, depending on the nature of the carcass; a bear can smell a carcass from literally miles away, and when a human stumbles on a grizzly that has taken possession of one, the grizzly reacts like a dog does when you take his food away, except that it's a grizzly. No matter what, a bear exhibiting aggressive craziness, like the Trout Lake grizzly, will at the very least be trapped and moved far away from people. And again, no matter where they are or what they are doing, people in Yellowstone are required to stay one hundred yards away from bears (and wolves, also), even when the bear appears level-headed and sensible.

We do not live in a safe era, where this matter is concerned. In all of North America, in the whole of the 1960s, there were no other fatal

The classic Yellowstone bear jam. GETTY IMAGES

grizzly attacks in the wild except the two on that single awful night in Glacier National Park. There were not many anywhere before that, and as noted, in Glacier, there were none at all in its whole history as a national park. Since then, as we will see, there have been quite a few, plenty of them in Greater Yellowstone. No one can say for sure why. In fact, the whole topic of "why" is cursed. If a topic of conversation were ever cursed, like the mummy's tomb is cursed, this one is it. Whenever there is an attack in Greater Yellowstone, or really anywhere, people jump on forums like Twitter/X and Facebook, and they express opinions about why it happened that they often should have thought twice about. In the wake of an attack, we reveal (I do it, too) that we cannot restrain our sudden expertise. After an airline crash, we are suddenly experts on "black box" flight recorders; after a bridge collapse, we are experts on stress fractures in I beams; after a bear attack, we are experts on the thoughts and feelings of bears.

In the course of this commentary, it is apparent that we cannot tolerate the thought that pure chance governs a good deal of what happens in the universe. We cannot stop ourselves from identifying the cause of events that happened far away, and that we learned about from a news website put together by a single twenty-year-old unpaid intern in New York City. We cannot stop ourselves from offering advice, especially on how to Be Safe. We cannot stop ourselves from mocking the dead for being inexpert, "foolhardy," and, the greatest sin of all, Unsafe. We cannot stop ourselves from revealing an intimate knowledge of what that animal was thinking. We cannot stop ourselves from predicting the future. And recently, we have often been so confident in our wisdom as to identify the cause as climate change.

The true explanation is that we can regularly never know why a given bear attack happened; as a result, we cannot routinely predict what they are going to do next, which is one reason they terrify us. The bears only explain themselves by implication and whatever evidence they leave. Sometimes, as we will see, the victim is clearly at fault, and sometimes, it is the bear. When it does look like the blame lies with the bear, a desperate effort follows to exonerate the "criminal." But of course, there is no crime. Police and lawyers speak of "criminal intent," without which they cannot place blame. Cause a death by accident, and it is manslaughter; kill in a fit of passion, and it is second degree murder; plan to kill, with intent, and it is murder in the first degree. But the bears, of course, know none of this. They are a good deal less to blame than Eve in the garden. At least she knew she had been warned. People who study bears have done what they can for decades now to identify ways to keep bears from attacking. Since motive is so often a mystery, they tend instead to address opportunity: Make it more difficult for the bear to hurt the human, and both sides are better off. But the truth is that we can never consistently know why the bear does what it does. The "motive," the "temptation," is forever opaque to us.

In all of North America, there were nine fatal attacks by grizzly bears in the 1970s. Among them was a deadly mauling in the woods near Old Faithful in 1972. The next most recent inside the park had been in 1942. The next most recent before that, again focusing on the park, had been in 1916, and before that, never—these things do not happen often. Because they are so rare, in addition to being so horrific, they tend to have an outsized impact when they do happen.

This one certainly did. On June 23, 1972, Harry Walker, who was drifting around the western United States with a friend who went by the name "Crow," entered Yellowstone National Park from the north. They appear to have been doing the kind of footloose and free-spirited exploration celebrated during that era; today, we would just call them "homeless." Because they entered the park by hitchhiking, they did not get the official warnings about bears that, then as now, are handed out at the entrance. Much was made of that in the aftermath of these events, including the protracted legal aftermath, but given what happened, would an official warning have made a difference?

The pair soon found themselves at Old Faithful, and what they did there was outlandish: They left the boardwalk at Grand Geyser and, somehow unnoticed, crossed into the forest behind the geyser and made a camp that was unauthorized and of course illegal. Leaving the boardwalk was and is illegal, too, and an unusually bad idea in an unusually active part of the Upper Geyser Basin. Crow told a ranger later that neither of them knew anything about Yellowstone. Back home, in Alabama, you could camp in parks if you wanted to, and they knew nothing about permits. They went back and forth for supplies, and somehow got away with it. During their first night, an animal entered the camp and could be heard rustling around near the tent. They found scat there afterward, and assumed it was an elk. When Crow described it later, it was pretty obvious that it came out of a bear. But, he said, "I don't know much about wild things."

On the night of June 24, they had drinks at the lounge in the Old Faithful Inn, one of the classier locales in the whole region, so one wonders where the money came from, although the implication that they were drunk may also need to be reconsidered, in light of the cost. Early Sunday morning, June 25, 1972, they surprised a grizzly that was feeding on food they had left out at the camp.

"Surprised" is the key word here. It is one of the most dangerous things a person can do to a bear. When you surprise a grizzly, their reaction is sudden and almost emotional—rash, we could call it. They often run away; when they stop and turn to examine the cause of the surprise, they do so (the theory states) after having covered the exact distance they need to feel safe. Sometimes, though, they run in the other direction, meaning they charge. It must be a defensive move, because from the bear's point of view, it drives the threat away. Nothing in North America is as frightening as a grizzly bear running right at you. Nothing. The human tends to crumple into a ball, or scramble up a tree, or run. The latter choice is regarded by experts as a mistake; the thought is that if you run away, you look like prey. You excite the bear. It remains, of course, a theory.

Crow and Walker were in no position for theorizing. They seem to have surprised the bear while the animal was feeding on the food they left out, and the bear reacted in defense of its meal (the reason a bear is also especially dangerous—really, *really* dangerous—when it has taken possession of a carcass). When they first reentered the forest where their camp was, they could not find it, and wandered lost in the dark. They were talking, idly, which is what you are supposed to do, although in this case it did not have the desired effect of warning the bear that it was not alone. They could see, it seems; they had a flashlight. The sky was dark the way it gets in the Northern Rocky Mountain wilderness; the village at Old Faithful does not produce much light, and it was cloudy enough that the rangers expected rain. The two at last stumbled on the path that led to the camp.

They first saw the grizzly when it was not more than five feet away, Crow recalled later. The bear was already charging, so the surprise was mutual. Crow reacted with pure instinct, diving to one side and rolling down a hill, in a cold and perfectly understandable panic.

At about this moment, the flashlight beam landed on the bear, and the bear now had a new target. It grabbed Walker and dragged him away. Crow heard Walker yelling, "Help me, Crow, help me!"

Crow yelled back, trying to make a dent in his own very great confusion: "Is there a bear there?"

He heard nothing more from Walker. He did not, however, hear nothing at all. "I heard more ruffling," he told a ranger later, "and it sounded like the bear was coming towards me."

So, he ran. For the next ten minutes or so, we do not know exactly what happened to him, because Crow was running, and screaming as he ran. He ran down the hill and through the thermal area. He should have stumbled into a hot spring and died horribly—there are, in that area, a great many to fall into—but he missed them all; during their visit to Yellowstone, the pair tested their luck to the greatest, most absurd extremes, and it went both ways for them. Crow kept running and screaming through the darkened geyser basin, its towers of steam rising all around him. He picked up the boardwalks, and they led him across the Firehole River. He had to pass through another open, undeveloped area. He at last burst into the lobby of the Old Faithful Inn, where he fell to the floor, gasping for air, and shouted to whoever happened to be present at that hour: "Bear! Bear! It's got my friend. Help him!"

There was little the shocked witnesses at the inn could do, or that *anyone* could do. They did try. There happened to be a ranger present, another stroke of luck. The ranger, Ken Reardon, approached with two other bystanders, who helped Crow to his feet and out to the ranger's car. They drove to the spot where, it appeared, Crow had crossed the Firehole, on the bridge (it is still there) over the river at Castle Geyser. Crow, the two others, and Reardon crossed the dark and empty geyser basin, and at first were not even armed. Reardon sent a trusted bystander back to retrieve a .38 revolver only when they reached Grand Geyser, right below the location of the camp. They even approached the hillside. A .38 is a policeman's sidearm and not really powerful enough to take on a charging grizzly bear; perhaps that is why it was first left behind. This little posse was not lacking in courage.

They *were* lacking in night-vision gear—it barely existed yet—and so retreated until they were joined by more rangers with weapons and flashlights. At 1:40 a.m., the rangers, carrying a rifle, a shotgun, and a handgun, began a search in the woods above the geyser basin. It was so dark that one of the rangers was surprised when it did not rain. Crow had, of course, given them only a vague description of where the events had happened. They found nothing, and gave up after a half hour. At 3:30 a.m., more help arrived, and another search began. The rangers, we should note, had been thrust into an impossible situation. Somewhere in the impenetrably dark and literally trackless forest above the Upper Geyser Basin, a man was lying, certainly injured and possibly, they had to assume, still alive and suffering. He could be bleeding to death, or on the verge of succumbing to internal injuries. In the same forest was an animal that had already proven itself to be as dangerous as an animal can be, and it would be a good guess that this creature would still be near its victim. The situation was as frightening as being an infantryman in combat in the war that was still going on in Vietnam. Still, they walked into the woods, quite literally looking for trouble.

For about two hours, they searched. At a little before 5:30 a.m., with Crow's help, they at last found the campsite. Some months later, one of the rangers, James Brady, told an official board of review that "he became a little emotionally disturbed and excited at that time, as you might expect." They now knew they were close. Just 150 feet away, down a hill and behind a rocky knoll, they found Walker, fairly quickly, given the circumstances; it was now 5:30 a.m., just about the time the horizon will have begun to glow with sunrise (it was close to the longest day of the year). But as events like these do, it all happened very fast. Crow, overwhelmed, "was escorted back to the Ranger Station where he could receive more medical care and attention."

The bear was not in the immediate area, and the rangers began to piece together what had happened. Crow had told them that it was Walker who was carrying the flashlight. They found it now in multiple pieces, about thirty feet from the campsite (one wonders: Did the bear target the source of the light that was suddenly, startlingly, annoyingly dazzling its eyes?). They then found a track on the ground leading from the flashlight a hundred feet to the location of the body, the trail marked, in part, by the contents of Walker's pockets. In the campsite, along the track, and on the bloody forest floor around Walker, there were no tracks or other sign that could tell the rangers what kind of bear, black or grizzly, had attacked. From the evidence at the scene, they could not even say it was a bear, although they worked on the assumption that it was. They photographed Walker, and worked up the area in the same way they might if it were a crime scene. They at last carried him off the hill and toward the Old Faithful Ranger Station. "About 7:00 a.m.," Ranger Brady said, "we were still at the site of the disaster when we heard someone shout 'bear' from down below us."

It was a pair of employees of the inn. They were on the boardwalk near Grand Geyser when they spotted a "big brown bear" at the edge of the woods. They yelled "Bear! Bear!" and got away fast. By the time rangers got there, the bear was gone.

Nerves were raw. The last fatal bear attack in Yellowstone National Park had happened here, at Old Faithful—but thirty years earlier. No one was used to it. No one, we can hope, will ever be used to it.

And the creature was still out there.

But not for long. The next day, Ranger Brady was joined by two of the park's wildlife experts, Doug Houston and Mary Meagher, who we will meet again later in this book. On the day of the attack, in the afternoon, they hiked to the site. Investigators had cleared all evidence that anything violent had happened here—except, of course, the blood. The three set out wire metal bear snares, simple and effective devices that are still in use. Each consists mainly of a can partly buried in the ground and open at the top; at the bottom is a bag full of some kind of smelly bait. The bear's nose leads it to the can. It reaches in and trips a lever that lassos its foot with a length of steel cable. When the bear pulls on the cable, it only tightens the lasso.

They placed the snares precisely where the incident had occurred. Two went in the campsite, and the third went where the body had been found. The next morning, not long after the sun came up, they returned.

When they reached the spot where the body had lain, the snare had been tripped, but was empty. Not so one of the snares in the campsite. Here, they found a grizzly, a female. Revealingly, they thought at first that she was enormous. Brady guessed she weighed 400 pounds, a size that would have made her about as large as female grizzlies get in Yellowstone. Later, she was discovered to weigh 232 pounds, almost half the size of the guess. To the investigators, we can see, she *looked* enormous. The snare was secured to a lodgepole pine, and she had figured out that she could get loose if she could get the cable off the tree. She could not get it over the top, so she tried the bottom, by digging—and had almost uprooted the tree by the time they got to her. Grizzlies are like that; they have those monster claws not to kill prey, mostly, but to dig—although these claws had been put to other uses.

The investigators heard noises in the forest, and wondered if it might be another bear. Brady said later that if this had been a black bear, or had been younger, or had clearly not been to blame, they would have drugged her, had a look, and released her. But no—this one was old; they could tell as she growled and snarled and revealed a mouth full of well-worn teeth. Brady knew that Walker's body had not been bitten much, and the canned food they found in the camp had been stomped rather than chewed open, so he suspected that he was looking for an old bear with what he called "worn dentures." She had also been tagged, and Brady recognized the number; he had trapped her before, here at Old Faithful, and a colleague had trapped her, too. She was, as Brady put it, "a former Craighead garbage bear" who had been hanging out at the Rabbit Creek dump, one of the open garbage dumps that John and Frank Craighead had used for their groundbreaking study of Yellowstone grizzlies—and once again, a garbage-habituated bear had made the worst kind of trouble. Feeding a grizzly garbage is clearly not a good idea, although this garbage bear would help change history. For now, a necropsy—an autopsy for bears—would almost certainly answer the question that was burning in the minds of the investigators and everyone else in Yellowstone, and millions of other people, too: Had they caught the bear that did it? So they shot her.

Looking back, it was clearly the correct thing to do. So eager was everyone for an answer that her body was taken that day to the state wildlife agency lab in Bozeman and examined before the day was over. In her digestive tract, the investigators found grass, which bears eat in great quantities all the time. They also found tinfoil, and caught on the edges of it, they found hair. Yes, they determined, it was human hair. There was more of it in her claws. They had caught the right bear.

She was old, this bear, grizzly number 1792 in the government records. Her teeth were ground down like . . . those of a twenty-year-old grizzly. They lead a harder life than we do. The canine teeth were broken, and all were worn down like a stick of chalk that is nearing the end of its useful life. However, for this time of the season, she had stored about the right amount of fat. She was not like the female that would later go on a rampage at Soda Butte Creek. She was reasonably healthy for a bear her age.

The bear had killed Walker fairly quickly; the autopsy made that clear. His body was examined by a doctor in Livingston, not far from where the bear ended up (Yellowstone National Park is a primitive place in many ways, and much of the business of life, including childbirth, is conducted in either Bozeman or Livingston, well north of park headquarters). The body was not a comfortable thing to look at. The nose had been "avulsed," the medical way of saying "torn off," leaving the nasal bones and nasal passages visible. The "entire skin surface," the doctor recorded, "shows multiple small lacerations and puncture wounds." As Ranger Brady had been told before he and the biologists made the decision to shoot the bear, there were few identifiable tooth marks. Walker had instead been worked over by those claws, with much blood and bruising around the marks.

But around the face and neck, there was also evidence that Walker had not suffered long. "The trachea," the doctor found, "has been extremely traumatized with multiple fractures and evidence of hemorrhage." The lungs were damaged also, and the bear had maimed his neck so badly that Walker died of suffocation. She appears to have used her claws and worn-down teeth as a kind of vice, with the trachea inside, and her strength was such that she was able to easily crush it.

The claw marks continued down the chest. The doctor, working top to bottom, recorded them and found that they disappeared at the abdomen—but not for any kind of "good" reason. They disappeared there because everything else did, too. The whole pelvic region was gone: "The lower abdomen for about an inch below the umbilicus is completely absent and all of the peritoneal tissue"—the lining of the abdominal cavity—had "been eaten away." Most of the large and small intestines were gone. There was not even any blood left in that area. Almost all of it was on the forest floor, or inside the bear, which had eaten about 25 percent of the body. The bear appears to have treated Walker as a kind of windfall. It had attacked because it was startled, but then, having been delivered by chance (the other side of that luck) with a meal, it proceeded to feast. The cause of death, finally, was due to "injuries, multiple and extreme, consistent with mauling."

The incident was thus wrapped up quickly. They had gotten the right bear, and almost immediately. The traveling public was as safe from dangerous animals as they ever could be in a place like Yellowstone. However, as the historian Lee Whittlesey has declared, "An entire book could be written on the Harry Walker case" (and in fact, since he originally wrote that sentence, at least one has: *Engineering Eden: The True Story of a Violent Death, a Trial, and the Fight over Controlling Nature*, by Jordan Fisher Smith). The Walker story was far from over, and that is one reason I have included it here. We live with its repercussions to this day. Along with the maulings in Glacier on that awful night in 1967, they changed forever the way we look at bears, and how we behave around them. Gone forever is the old laissez-faire indifference that told us not to worry about these animals that were so charmingly goofy that they could hardly be a threat. This attitude was inadequate in a way that was almost suicidal.

What had gone wrong that night? More than you may think. On November 6 of that year, the official NPS Board of Review met at headquarters to figure out what had happened, and when all the evidence was assembled, it was more than clear (it was already becoming so the morning after) that Walker and Crow had done just about everything wrong. As Crow had told the rangers, they knew nothing about Yellowstone and bears, but it is almost as if they had been reading a mirror-universe version of Professor Herrero's book, written by his evil twin (and so titled *Bear Attacks: Here's How to Provoke Them*). One of the most important things campers can do to stay safe overnight in bear country is to keep a clean camp, with all food trussed up and hanging high in the air, but Walker's campsite was like a banquet for the animals.

Whittlesey, who was a junior tour guide at the time, recalls a ranger telling him the camp "was as dirty as any he had ever seen." The board of review asked one of the investigating rangers, Gerald Tays, what he had found there. Had the bear been feeding in the campsite? "Oh, yes," Tays responded. "There was considerable food left open in the campsite to begin with. We found a pot of rice and other already cooked food hanging in a tree. Also, on the ground near the tent was the remainder of freshly cooked food that had been the evening's supper. There had been, to the best of my remembrance, a jar of jelly which had been broken open, [and] a flour material spread from one area to the other which the bear had gotten into."

It gets worse. Ranger Brady told the board that Walker and Crow had set up camp next to a garbage bear highway: "The area in the past has been frequented by bears and there are bear signs all along the ridge, just above the geyser basin," by which he meant signs indicating that bears had been around, like droppings or diggings. "If you notice back behind, there is a long ridge that actually ties in with the Rabbit Creek area. It's heavily traveled by black and grizzly bears. You could find signs up there right now." The Rabbit Creek dump had been closed two years earlier, but its former denizens, like bear 1792, were still in the area. The risks added up and synchronized with one another. The bear was acting like a garbage-habituated animal will; she was comfortable being around things human, drawing her dangerously close to people. She was also acting as *any* grizzly will. In the camp, she had found a generous food source, and when grizzlies nail down food in this quantity, they use lethal force to defend it.

In her book *Taken by Bear in Yellowstone*, Kathleen Snow itemized the four most glaring errors Walker and Crow made that night:

- They were not camped in a designated campground.
- They camped in an area heavily used by bears.
- Their food was accessible to a bear.
- The victim and his companion were hiking at night.

She believes the first two factors to be less important than the last two, but it is also true that if the pair had stayed in the Park Service campground at Madison, sixteen miles to the north, or any of the other official campgrounds, you would not be reading this, because the attack would never have happened. There is safety in numbers, although as Crow told the investigating rangers later, the two were looking to get away from crowds ("If I wanted to go camping with a bunch of people," he said, "I could just go down on the river bank in Alabama"). So, they left the safety of the crowd and, although they could not have known it, camped in a spot that greatly increased the likelihood of a bear encounter.

They were, honestly, largely to blame. Naturally, Walker's family did not see it that way. They sued the National Park Service and were awarded $87,417.67. The Park Service got an appellate court to reverse the decision, and in the end the Walkers never collected any money.

The attack, however, had an enormous impact beyond the personal blow to the Walker family. Whittlesey summarizes the effect it had on the park and its bears. Although the attacking bear was a grizzly, and so had nothing to do with the spectacle of roadside feeding involving the black bears that had gone on nearly since the park was founded, bear management was shaken up from top to bottom. Both species were affected.

> *The Walker incident served as the final nail in the coffin of bears being allowed to feed along Yellowstone roadsides. The last open-pit garbage*

dump in the park had been closed in 1970 at Trout Creek, and now the NPS stepped up the transporting they had begun in order to get all roadside beggar bears off their diets of human food. Normally reclusive bears can lose their fear of man when human foods are involved. A more natural bear is a less dangerous bear, and the NPS knew this. The dual attacks in Glacier National Park in 1967, as related in the book Night of the Grizzlies, *had already cemented that fact in the minds of managers there, and now Yellowstone had its own version.*

The change did not happen instantly. In fact, it took a long time. To this day, people come to the park and ask where they have all gone; older people who visited in the 1950s or 1960s remember the roadside circus of their youth, and demand to know what the government has done with the bears. They are still out there, more than there were before. As noted, your best chance to spot one will be up north, near Tower-Roosevelt or along the Northeast Entrance Road. When you do see one, it will likely be a truly wild animal.

It is wild, yes. Therefore, it is dangerous.

Closing the Yellowstone dumps and running the bears away from the Grand Loop Road has not, in fact, made us safer from bear attack. In the half century plus since these changes were implemented, bears have attacked people in Yellowstone National Park over and over. In Greater Yellowstone as a whole, it has happened in a way that has almost been routine. It is not a common way to die, but the danger is real.

Let us focus just on the park, for the sake of brevity. In 1984, a young Swiss woman named Brigitta Fredenhagen hiked alone into the Pelican Valley area in the central part of the park, where she spent the night of July 30. Fredenhagen was in many ways typical of young European visitors to the park: Sophisticated, adventurous, and speaking nearly flawless English, she appears to have perfectly understood the danger the bears posed. Indeed, she did only one thing wrong: Entering a part of the park that grizzlies enjoy a great deal, she went alone. Her brother reported her missing on the evening of July 31, and the next morning, ranger Mark Marschall found her campsite. He knew something was wrong when his horse shied and would not approach closer than twenty yards. Approaching her tent on foot, he saw that its fabric was torn, and noted, on the ground, pieces of scalp, hair, muscle, and bone. He called for backup, and after two more rangers were helicoptered in, they continued the search. At the end of a trail of bloody clothing and human tissue, they found Brigitta, partially consumed. The bear had apparently attacked her in her sleeping bag, the kind of attack that normally happens when a bear has settled on a human specifically as a meal. On this occasion, the bear, presumably a grizzly, got away with it. The animal was never identified.

William Tesinsky was also alone in the backcountry when he became the next victim inside the park. Tesinsky was an auto mechanic who had some talent as a wildlife photographer, and he had experienced just enough success selling his photographs that he was, in the autumn of 1986, trying to establish himself as a professional. He had never photographed a grizzly, and the gap in his portfolio weighed on him. He was like the freelance photographers who, during the Vietnam War, wanted to crash into the news profession so badly that they took risks they should not have taken. I sympathize; I know from experience that when you are an auto mechanic in rural Montana, breaking into a creative field is not easy. Risks are pretty much required.

Dan Sholly, the chief law enforcement ranger at the time (and father of the present park superintendent, Cam Sholly), investigated the matter. "Sometime on a cold afternoon three days before his remains were discovered," he recalled later, "the beast of his dreams had appeared as if by magic in a meadow. Tesinsky happened to be driving by in his 1963 blue Chevrolet Impala. And so Tesinsky left his car in a pullout and began stalking the distant brown furry form—on foot, alone, camera and tripod in hand, his whereabouts unknown to anyone." At some point, well away from the road, the bear noticed him and charged. The bear, a sow numbered 59, treated Tesinsky as bear 1792 had treated Harry Walker, as a windfall, much appreciated here in October, with winter coming fast. The authorities finally started looking for Tesinsky when the Impala had been in that turnout for a period of days, and knew they had found him when they spotted 59 walking around with a human leg in her mouth, a tennis shoe on the foot. They killed the bear and retrieved the body, which was missing its upper half, all of it. One of the rangers thought that it looked as if the body had been cut in two with a chainsaw.

There followed, after 1986, a long break from the ursine mayhem. That changed on July 6, 2011. Brian and Marylyn Matayoshi were hiking on the Wapiti Lake Trail near the Grand Canyon of the Yellowstone when, in company with another hiker, they encountered a sight both endearing and terrifying, at least when on foot in the backcountry: a mother grizzly with two cubs. Here is one of Yellowstone's truly dangerous phenomena, about as explosive as the volcano itself. The Matayoshis photographed the animals, then moved on. It is not clear whether they were the same bears, but a mother grizzly and two cubs appeared later, when they were alone. The animals were dangerously close, only a hundred yards away.

The couple turned and walked in the other direction. The Matayoshis were not Yellowstone newcomers; this was their fourth trip to the park. They knew what they were up against here, but they had omitted bringing a crucial item, one they should have known about. Here in 2011, we have entered the era of bear spray, which did not exist in 1986, or at least not in a form available to the general public. They likely were feeling its absence as they hurried away, and then Marylyn saw the mother bear. She seemed to "pop up," as Marylyn later put it, and then she charged.

Brian yelled, "Run!" Again, this is universally regarded as a mistake: The idea is that the bear will think of you as prey if you act that way. More important than reading the bear's mind here is the undeniable fact that bears can run much, much faster than humans. Marylyn heard Brian cry out, and turned to see the bear hit Brian and knock him to the ground. Even the cubs were charging now. Marylyn found a fallen tree that offered a trace of cover. She threw herself flat, but peeking up, she saw the bear staring right at her. She fell face-first to the ground again with her arms over her head in the manner of a soldier taking cover from an artillery shell. Then the bear was on her, picking her up into the air by her backpack alone and then, blessedly, dropping her and disappearing. The attack had lasted perhaps a minute. The bear had reacted defensively, and had stopped the mauling when Brian fell silent. After the bear was gone, Marylyn ran to Brian. His leg was bleeding freely; she tried to use a jacket as a tourniquet, but the bear had bitten down and laid open

Brian's femoral artery. No autopsy was conducted (it was pretty obvious that a bear was responsible), but the injury to his artery would have been enough to kill him. Furthermore, when the bear had taken him down originally, she did it with a blow to the chest that a doctor later speculated could have been fatal in and of itself. That is how bears often kill: When they want to take down an elk or moose, they just hit it. A single blow can break an elk's neck.

As part of the investigation, a ranger recovered a pair of broken sunglasses with six or eight hairs from a bear on them, obviously from the bear who was responsible. The hair was saved as evidence; we had also entered an era in which single hairs could yield enough DNA to identify a culprit, human or otherwise. The investigators could never have guessed that the hair would play a role in yet another investigation.

Seven weeks later, on August 24, 2011, John Lawrence Wallace arrived in the park alone on a visit from his home in Michigan. He was a gentle, thoughtful man who ran a library at home, and so in many ways was typical of Yellowstone's most avid fans. He referred to the park as "heaven." He stayed in the campground at Canyon, and early on the morning of the 25th, he parked at the Mary Mountain trailhead in nearby Hayden Valley and started hiking. If he was planning on hiking the whole trail, it would take him through one of the genuinely remote parts of the region, to emerge at Fountain Flats, twenty miles away. The attack seven weeks earlier had happened near Canyon, not far from this trail, but if he was thinking about it at this date, he might be forgiven for seeing a certain level of protection in the landscape itself. The Wapiti Lake Trail is in the Canyon area—but on the other side of the Yellowstone River. There is of course a bridge, the Chittenden Bridge, to carry the road, and bears can swim, but it is quite a barrier.

No matter. Wallace was not carrying bear spray, which—given that he was alone in some of the best grizzly bear habitat in the park, and maybe even the world—was by this point in history really inexplicable (the stuff, we should at least note, is expensive). It made less sense given that he passed a dead bison where, three days earlier, a hiker had seen nine different grizzlies (*nine*) feeding on the carcass. A second dead bison had, rangers later found, sixteen daybeds around it, left by bears. The next morning another park visitor and his daughter were on the trail, six miles from the trailhead, when the daughter noticed birds circling a daypack and water bottle. Approaching, she saw a thoroughly violated body and much blood; it was Wallace, who had in fact died of what doctors call exsanguination—he had literally bled to death. The body was at the end of a trail of blood and bloody bear footprints, belonging to an adult grizzly and probably one cub. The bear had attacked and quickly dispatched him, and then "cached" him: The body had been buried, but only partly, the way bears do, and had been partly consumed. If anyone had been interested in sitting and waiting, the bear would have returned. DNA from scat piles near the body matched that in the hair from the sunglasses found after the Matayoshi attack, although given that a small army of bears was in the area, another may have killed Wallace, and the animal the rangers came to call the "Wapiti sow" made a meal of what, again, was windfall. We have no witnesses, or no *human* witnesses. The authorities later caught the bear and its cubs, and killed the mother. The only place for her, by that point, would have been a zoo.

I should note that all these attacks remain the subject of animated conversation in the extended park community. In private, the evaluations of what happened are often unforgiving, nasty, even vicious. Among the cases I describe in this chapter, for instance, is one in which the victims are described—I have heard this more than once—as "the bums." You hear the word "foolish" all the time, even though it is not possible to take all the precautions recommended by the "experts," the same people, often, who are always reading the bear's thoughts.

As I wrote the first draft of this chapter during summer 2024, the town of Gardiner, Montana, was being very nearly terrorized by a bear, obviously a grizzly, who showed up early in the season and made himself at home. We keep track of these things through our best source of local news, the Gardiner/Mammoth and the Mammoth/Gardiner community message boards on Facebook; one is supposed to be conservative, the other leftist, although I cannot tell them apart by ideology alone. One of my neighbors in town, early in the whole affair, posted photos of bear scat taken in the dirt alley directly behind my house. The bear had been eating garbage, and one of the single loudest alarms a bear can send attesting that he is dangerous is bits of garbage in his bowel movements. These were so full of plastic that one was, quite literally, gift-wrapped.

We kept track of his doings by means of webcam footage, posted on one of the two Facebook pages. He broke into a truck with consummate skill. He broke into the coffee shop at the Sinclair station and dined in. A bear's nose will tell it that coffee has no calories in it—he was after cream and the like—although it is easy to imagine a bear developing a taste for caffeinated beverages, after which things would get even more interesting. He tore the window out of a house. He was filmed walking down a street carrying a bag of garbage in his mouth like Charlie Brown's dog Snoopy carrying his food bowl. He ignored the bear trap left in the middle of his nightly rounds, apparently understanding perfectly what it was.

In the end, officials with Montana Fish, Wildlife & Parks shot the bear, a male grizzly; there seemed to be no other choice. But in the meantime, were we supposed to follow all the Stay Safe in Bear Country rules—in town? One rule everyone pretends to follow, even though it is impossible to do so, is the one stating that you may not enter the backcountry safely in a group smaller than four adults, lest you be called foolish after you are killed and eaten. Should we have assembled into groups of four to go to the c-store, or the house next door, or—given the way things seemed to be going—to the bathroom?

Still, a number of the fatal incidents I have described should never have happened. In every case except the Walker mauling, after which Crow freely admitted that they knew nothing about Yellowstone or bears, the victim really ought to have taken some bit of precaution that would have headed the whole nightmare off. The last of our fatal park maulings, the 2015 Crosby incident, is one of those about which people have strong opinions.

Because Lance Crosby was as well-informed a hiker as we are likely to come across. Of the victims we have looked at in this chapter, Fredenhagen was from Switzerland, Tesinsky from Great Falls, Montana, the Matayoshis from Torrance, California, and Wallace from Chassell, Michigan. Crosby was different because he actually worked in the park, and had for years. Nor was he a typical park employee, who tend to be footloose twenty-something youths who are in Yellowstone for a lark, or footloose thirty-something felons who are in Yellowstone to keep one step ahead of the law. They often know no more about bears than the tourists. Crosby was sixty-three and married with two children. He was a registered nurse with a master's degree and worked for Medcor, the organization that ran the medical clinics in the park; he worked at the little hospital next to the Lake Hotel. When he set out on the local Elephant Back Trail on August 6, 2015, he was very nearly hiking in his backyard.

If August 6 rings a bell, it should. That day was, weirdly, the seventieth anniversary of the atomic bombing of Hiroshima. The coincidence is meaningless, except to remind us that the narratives of disaster in this chapter are not so enormous in the history of the world, but are instead merely stories of bears killing and eating as they normally do. Much worse things can happen . . . much worse, unless you are involved. A goal of this chapter is to suggest ways you can avoid becoming involved. The line from *Jurassic Park* comes up in Yellowstone more often than elsewhere: "They just do what they do." We need to avoid getting in the way.

As he hiked over a local feature, Elephant Back Mountain, Crosby was not exactly in the howling wilderness, even though he was off the trail. For her book *Taken by Bear in Yellowstone*, Kathleen Snow spoke to one of Yellowstone's bear experts, Kerry Gunther, about the terrain. "This is not remote wilderness," he said. "Most use of this off-trail ridge area is made by local people who live there, who are out walking for exercise or exploring. The site of the attack is less than a half mile from the Grand Loop Road, and one mile from the nearest house." Perhaps because he was hiking in his own backyard, he made the two mistakes we have seen repeatedly in these cases: He was hiking alone, and he was not carrying bear spray. His coworkers at the Lake hospital reported him missing the next day, and a ranger found the body that afternoon. As we have also seen repeatedly, the bear was a mother with cubs; she had killed Crosby, consumed part of the body, then cached the rest. The authorities killed the sow, and the cubs went to a zoo in Toledo, Ohio. The NPS came in for a torrent of abuse for killing the mother, and we should note here that they do not like to do it: The park superintendent, Dan Wenk, meant it when he said in a news release, "The decision to euthanize a bear is one that we do not take lightly." The thought is that, given how uncertain a future the bears face, they need as many human friends as possible; having a bear repeatedly kill people when the government might have stopped it is bad publicity the animals do not need. The people do not need it, either, but they are also thinking of the animals. The people involved are always the park bear experts, who have devoted their lives to the welfare of *Ursus arctos horribilis*. As Kerry Gunther put it, in his interview with Kathleen Snow, "Society's willingness to protect grizzly bears and the large tracts of habitat necessary for grizzly bears to survive might be significantly diminished if bears were allowed to regularly kill and consume people."

Focusing on the park itself, the Crosby incident was the last fatal attack; as I write this, we are going on a decade without a bear-caused human death inside Yellowstone National Park—although as noted, the violence has continued in the surrounding national forests, where in fact, it never seems to stop. Looking at the dates, it may feel like the park is "due" for a fatal attack. You should actually resist thinking this, because it is a variant on what philosophers call the "gambler's fallacy," the common error we make when, for example, a flipped coin lands on heads five times in a row. On the next flip, you think that the odds of it landing heads, of the streak continuing, are indescribably enormous, right? Wrong: Whenever you flip the coin, the chance is always fifty-fifty. The coin does not know it has had a long run. Neither do the bears.

It is up to us to keep the run going. You would not want to rely on the bears to do so, although heaven knows people have tried it. For an example, watch the film *Grizzly Man* sometime, Werner Herzog's classic documentary about Timothy Treadwell and his girlfriend Amie Huguenard, the conservationist couple who befriended bears in Alaska, or thought they were doing so, and were eventually killed by them; they seem to have genuinely believed that the bears were their best friends. They were not—no, not at all. Bears are one of the only large animals that will kill you on purpose, and do so in order to eat you. Other animals in the region are capable of doing so, but only bears do it regularly. Even when they kill someone "by accident," as when a hiker startles them and they attack, they still keep the body. Notice how many times bears have partly eaten the body of a backcountry visitor they may have killed defensively, and gone so far as to cache it in the same way they would any other animal, burying it partway with the intent of returning. Bears might or might not be the most dangerous animals in the park—I think the most dangerous is another, and they will come up later in this book. They are, however, surely the most frightening. And they should be.

We are capable of deceiving ourselves, with bears. An individual like Hank the Tank will have what amounts to a fan club, and a large one. Videos of their antics charm people by the tens of millions. It comes down to this: We find them cute. They appear in cartoons and online memes in vast numbers, rivaling dogs and cats as subjects. At some level, we think that this animal, model for the teddy bear, could not be terrifyingly vicious.

They certainly can be. You should not, however, allow yourself to be so frightened of them that you will not leave the car or the hotel room. Some people do that, getting so worked up that they ruin their visit. A few precautions will keep you at least relatively safe. No one in Yellowstone is ever perfectly safe.

Happily, we have one great invention that levels the playing field when the two species meet: bear spray.

We covered the basics about spray earlier, but the subject needs more attention, because it is a serious weapon. One piece of advice that you do not usually get is that you should try to test-fire the bear spray first, before going on any kind of serious hike in the Yellowstone wilderness. There are a number of surprises, and you should encounter them first in a practice session, and not in a life-and-death encounter.

Do not use the canister you intend to carry in the wild; they do not last that long. Canisters have expiration dates printed on the side,

The revolutionary innovation: bear spray.

and a neat trick is to find an expired canister. Talk to some locals and offer them five bucks for one; they often have them stacked like artillery ammunition in a closet. Then go far away from people, maybe out along the edge of one of the national forests, and create a firing range for yourself by finding a target that is about the size of a bear's head. Walk about fifteen feet *upwind*—make absolutely sure that the wind will carry the spray away from you. It is incredibly, unbelievably awful stuff. Put the bear spray on your hip, preferably secured to a belt, which is the place you should wear it in the backcountry. You have to be able to get at it fast, if it comes to that. You need to think militarily on this matter.

You will draw and fire in five motions. Wear the bear spray on the side your hand favors, right side for right-handed people, left side for left. Double-check and make sure anyone with you is behind you, and always treat the bear spray as if it were a small-caliber gun. If you are right-handed (lefties, just reverse all this), first pull the spray quickly out of its holster and upward with your right hand. Second, grab the body of the canister with your left hand. Third, pull the safety backward and free with your right hand (you can eliminate a step by thumbing the safety loose, but that requires dexterity that might escape you when a bear is

Ready.

Fire. This is the two-handed technique, best for careful aiming.

charging). Fourth, put your right index finger through the thing that looks like a trigger housing. The actual trigger is on top and was made ready to fire when you removed the safety. Finally, in one motion, aim at the target and, with your right thumb, press down on the trigger.

And prepare to be surprised, and maybe even shocked. The canister makes a bark and spits out a cone of spray that will reach thirty feet, and even farther. The natural analogy is with those little mace canisters that go on a keychain, but there is no comparison.

So, we can defend ourselves now. Before the development of bear spray, the only defenses were playing dead or climbing a tree. The thought of playing dead is one that we all find impossible to get used to, and climbing a tree is so difficult in Yellowstone, given the tree species there, that rangers are rethinking that piece of advice. People have been carrying firearms as a defense against bears since the era of the frontier, but here is an entirely unpublicized problem I have noted regularly whenever a hunter is mauled. Often, a bear makes a false charge to frighten off a human it regards as a threat. Killing a bear under such circumstances requires an almost unbelievably icy nerve, because the shot has to hit a minuscule patch under the bear's chin for the bullet to pass through the heart and so kill the bear on the first shot, which is the way it must be done—and this, while a mortal threat is hurtling toward the shooter at thirty-five miles per hour, making noises that are uniquely horrible. The line from *Blazing Saddles*—"If you shoot him, you'll just make him mad"—ceases to be a joke. The hunter who shoots at a charging bear and only wounds the animal turns a false charge into an attack.

Bear spray was developed in the 1980s by bear biologists Chuck Jonkel and Carrie Hunt. When Jonkel died at age eighty-five in 2016, *High Country News* published a tribute written by Ben Long, a reporter who has covered the "bear beat" in Montana. He thinks that bear spray has done more than just prevent attacks:

> *Bear spray has proven itself over and over again. I've interviewed dozens of hikers, hunters and berry-pickers who have repelled troublesome bruins with pepper spray. Scientific studies have also demonstrated the effectiveness of spray, showing it works better than firearms and has the added benefit of leaving the bear alive.*
>
> *Recently, I discussed this with Gary Moses, a career ranger in Yellowstone and Glacier national parks who now is a representative for Counter Assault [a popular brand of spray]. Moses notes that, a couple decades ago, two or three visitors were mauled by grizzly bears every hiking season in Glacier National Park. In recent years, the number of maulings has dropped, even as the number of visitors has climbed substantially, and the number of bears has edged up as well.*
>
> *Moses has no hard data to point to, but he believes that the difference is pepper spray, which has become standard equipment for rangers and hikers alike. Now that bear spray is ubiquitous in national parks, hikers are better able to defend themselves from troublesome bears.*
>
> *But that's just the start of it, Moses believes. Grizzly bears are highly intelligent, and cubs learn life lessons from their mothers. Moses thinks that more and more Glacier grizzlies have learned to avoid people after being hit with pepper spray. Those bears go on to teach their cubs to give people a wider berth. It's just a hypothesis, but makes sense to me.*

And to me as well.

Lessons Learned

I have devoted a substantial portion of this book to the grizzlies because they deserve it. They are singularly terrifying. We looked at the stories of the people who have been attacked and killed because that is one way you can learn to avoid ending up in their blood-soaked shoes, by finding out what they did wrong. The attacks we looked at also led us, starting with that night in Glacier and continuing to that night in the woods above Old Faithful, into the modern bear era. Fatal attacks have a way of changing the way we do things. Jack Olsen, who wrote *Night of the Grizzlies*, thought that the dual Glacier attacks in 1967 had doomed the grizzlies. He was writing in 1969, and he thought that in the future, they would be hunted everywhere to extinction in response to this shocking proof of their volatility. He was wrong, happily, but we did change the rules. The garbage is mostly under control, and people take all kinds of sensible precautions now. That first attack we looked at, in the campground at Soda Butte Creek, reminds us just how surprising bears can be, so sensible precautions are a permanent necessity.

I wanted to take a long look at the bears because I also wanted to set the right tone. Grizzlies are what makes Yellowstone what it is, for deeper reasons than are usually publicized, and these reasons are central to everything in this book. They have to do with finding a real alternative to the fake life. As we have seen, and will continue to see, Yellowstone puts you on alert. You cannot help but be fully alive here, because your instincts will recognize that you are sharing the environment with these mammalian main battle tanks. When you are on foot in Yellowstone, and away from the pavement, you are more or less perpetually jazzed. It is one of our major—mostly unpublicized—attractions.

Do bring the bear spray, though. It has not made Yellowstone safe, but there is no way to make Yellowstone safe, and it would be stupid to even try. There are so very *many* creatures here who will hurt you if you let them. Even taking a long look at the bears is only a beginning.

Let us now look at some of the many, many other ways Yellowstone can get the better of you.

Chapter Three

By Their Fruits Ye Shall Know Them

Plants and Pathogens

One thing about bears: They are big. When they behave in a threatening way, they leave you no doubt that you are in trouble, in part because they are so superlatively *large*. You cannot ignore a seven-hundred-pound animal crashing through the forest and heading right for you. It is unsubtle and rather direct.

Other dangers are more sly. Some of the creatures in Yellowstone that will do you in are microscopically small. They can be said to lurk—except that "lurking" implies intent, and these creatures are not capable of intent. They are perfectly mindless. They, too, just do what they do.

Even some of the plants will get you. In some situations, the plants are not quite to blame. Falling trees sound like a disarmingly silly threat, the kind of danger stalking characters in old cartoons, but talk to lumbermen about it sometime. They will not be so dismissive. They have a name for heavy branches on a tree that break off and remain hanging in the canopy, ready to come down unannounced: widow-makers. There have even been times in history when falling trees killed people in large numbers, like an epidemic. During the Second World War, in the South Pacific, the rain forest itself could become a threat. The drenching tropical rain, combined with shaking from various kinds of bombardment and even just the natural earthquakes that happen on volcanic islands, loosened the roots of trees often well over a hundred feet tall. In places, they were as deadly as the combat itself. In the fighting on Cape Gloucester, on the island of New Britain, fifty US Marines were injured by falling trees, about half fatally. It was hardly a trivial issue.

There have been four deaths inside the park, and possibly five, from falling trees, "possibly" because some of the early records are vague—and here again, I have to violate that promise that no one dies in this book, but thoroughness requires it. We do not know how many people have been hurt. A number of the deaths occurred during the early era of the park, when a great many trees came down as part of various construction projects. More recently, a child was killed in the old, now-removed cabin area at Fishing Bridge in 1936, and a man was killed in 1966 in the campground at Lewis Lake by a falling lodgepole pine that had been rotted out by a rust canker, a kind of infectious disease caused by a parasite. The family of the victim sued successfully, and it is another of those cases in which the Park Service has been found liable for a natural process it could scarcely control. If you find your campground has no trees in it, that case was probably to blame. The trees were cut down for safety's sake.

Trees do not hit people "on purpose"; it is not part of their natural defense. The park, however, does have poisonous plants—that is, plants that will hurt you very much "on purpose."

Mushrooms found growing a few miles north of Mammoth. They look delicious, but taking a bite would be a gamble.

Two plants come up when the subject of deadly herbage in Yellowstone National Park is under discussion: water hemlock (*Cicuta douglasii*) and the unsubtly named death camas (*Toxicoscordion venenosum*). There are also six mushrooms that will kill you, and they have names that are even more unsubtle, if kind of poetic: death cap, destroying angel, deadly conocybe, deadly cort, deadly galerina, and false morel. Mushrooms have never killed anyone in the park (that we know of), but again, the park has only been a "park" for a small fraction of its existence. The rest of that time is mostly lost to us.

One fear people have arriving in Yellowstone from other parts of the country, especially when there are kids and pets in tow, is of the classic plants we worry about at summer camp: poison ivy, poison oak, and poison sumac. The latter two, oak and sumac, do not grow in our part of the Rockies, either in the park or in Greater Yellowstone. Poison ivy, a plant of various species in the genus *Toxicodendron*, is a different matter, although not a major cause of worry. Howie Wolke, an outfitter who works in the area, writes of his own experience with it: "Over the years, I have hiked much of the Teton Range and have never seen poison ivy there. I have, however, read that it does occur at the foot of the mountains along the western shore of Jackson Lake, a relatively low-elevation area with plenty of moisture." I have read that, too, and have also never seen it there. "By contrast, in Yellowstone I have seen poison ivy—but only in one location: a low-elevation south-facing slope with spring-fed moisture near the confluence of Hellroaring Creek and the Yellowstone River. And that's it. Nowhere else. Yet I have read that poison ivy is absent from Yellowstone, which obviously isn't true." Hellroaring Creek is well into the backcountry, so you will probably not run across it—although it is on a trail, one served by a suspension bridge over the Yellowstone River that is one of the man-made sights of the park, so it is not as if it were on the moon.

One intriguing "plant" that belongs in this discussion is not a plant at all. It is easy to spot, hanging from the trunks and branches of trees, because the color is so distinctive and so in-your-face: lime green, roughly, or Day-Glo green, or maybe chartreuse. Look at it closely (it will not hurt you, as long as you keep it out of your mouth), and you will see that it is branched like a piece of coral. It is wolf lichen, *Letharia vulpina*, the "vulpina" from the Latin word for "fox." The names come from the ancient purpose to which this widespread lichen was put, as the source of a poison used to kill wolves and foxes. One particularly horrible early recipe called for wolf lichen mixed with broken glass, with the glass intended to allow more poison into the victim's system. The poison, vulpinic acid, is part of what gives wolf lichen its bright color, which seems to serve as a warning to any creature that might be thinking of eating it. It is not terribly poisonous, but you would not want to mix it into a salad (it is said, by the way, to taste like aspirin). As with so many things in nature, it cuts both ways: It has antibacterial properties and may protect skin from ultraviolet light. Still, keep it away from small children.

Wolf lichen, in a forest on the northern range.

Of the two deadly plants, death camas has never killed anyone in the park—again, that we know of. Death camas is a threat to the ranching business, since cows will eat it, and will in fact poison themselves to death eating it. The poison is concentrated in the bulb and mature leaves, and a cow eating as little as a half-pound may have thereby doomed itself. Human poisonings are not common, but when they do happen, they are sensational. A classic reference book on this topic, *Common Poisonous Plants and Mushrooms of North America*, details the symptoms an unlucky sufferer may endure, including "excessive watering of the mouth, burning followed by numbness of the lips and mouth, thirst, headache, dizziness, nausea, stomach pain, persistent vomiting, diarrhea, muscular weakness, confusion, slow and irregular heartbeat, low blood pressure, subnormal temperature, and in severe cases, difficulty in breathing, convulsions, coma, and death." It must be exhausting.

The greatest danger may come from mistaking death camas for wild onion, which can be found in many places in Greater Yellowstone and is delicious—and it is perfectly legal to eat it. People have the sense that such an activity must be forbidden somehow, but it is not; the reasoning is that the National Park Service is in the business of preserving local history and traditions, and foraging is one activity that has gone on in Greater Yellowstone since the ice retreated. Above the ground, death camas and wild onion actually *do not* look similar. The flowers on a death camas appear as a vertical stack of blooms growing up and down a central stalk, while wild onion is like a domestic onion allowed to go to seed: The flowers are in a single mass on top of the stalk. There are a number of other differences, but a crucial one is in the bulbs. They appear identical, which is really bad, because as few as two of the bulbs will kill a full-size human. They lack, however, that distinctive smell. Probably everyone alive, over a certain age, knows that smell, the unmistakable aroma—part noxious, part delicious—of a freshly cut onion. A death camas bulb does not have it at all.

The other dangerous plant in the park, water hemlock, definitely *has* killed. This plant is thought to be the single most poisonous plant in the northern temperate part of the earth. It does not grow beneath the waves, as the name seems to suggest, but does have an association with water, growing in wet meadows and along streambanks, as well as other damp places, which Yellowstone has in great abundance. It can be two feet tall, or well over six. It puts forth small white flowers that grow in—we need the botanical terms—a compound umbel, also called an umbel of umbels, which is actually easy to picture. When flowers grow at the end of multiple little stalks branching away from the main stalk, that is an umbel, a term that shares its word-root with the ribs that branch away from the shaft of an umbrella. In the compound version of an umbel, the feat is duplicated: At the end of the small stalks, another set of stalks, smaller still, radiate away with the flowers at the tips, like bomblets thrown out of the main bomb in a firework display.

Water hemlock roots make tubers, and this is where people get in trouble: The roots look edible. The plant can be mistaken for a shockingly wide range of edible plants: water parsnip, wild parsnip, wild carrot, celery, ginseng, and probably others. It is said to smell like fresh turnips, and tastes sweet. The poison itself is called cicutoxin, from the name for the genus, *Cicuta*; it is what doctors call a neurotoxin, a poison that, once ingested, goes to work on the central nervous system, creating the characteristic symptom of cicutoxin poisoning: violent convulsions. Victims will first, and quickly, become nauseous, and may vomit enough of the root up to save themselves; vomiting is unpleasant, but that is why we do it. A severe respiratory distress follows, with cyanosis: The victim literally turns blue. Perversely, the person also flushes, and so actually turns red and blue. Sweating, salivation, and delirium develop. According to the Centers for Disease Control (CDC): "In fatal poisonings, severe seizures occur after the initial symptoms, and death results usually from status epilepticus"—that is, either a seizure that is continuous, or a series of seizures that happen so close together in time that they might as well be the same one. "The case-fatality rate for poisonings reported from 1900 through 1975 was 30%." That is about the same rate as smallpox, before the vaccine.

The medical literature is filled with some amazing cases involving cicutoxin. In one, a child fashioned a whistle out of a water hemlock stem, and the mere contact of the stem with his lips was enough to be fatal. In another, a family rubbed a salve made from water hemlock on their skin; two members of the family died, so the toxin can be absorbed that way, too. One case was reported by the CDC: In 1992, a pair of brothers were foraging in the woods in Maine and found what they thought was wild ginseng. One brother had three bites of the root, the other, one bite, of what was actually a close relative of the water hemlock we have in Yellowstone. The brother who took three bites died in the emergency department of a local hospital three hours later; the brother who had one bite was treated as a precaution and survived—after going into a delirium, with seizures.

As you will have guessed, it is easy to make the fatal mistake of eating this stuff, as our collective experience in Yellowstone shows. One of the victims of fatal poisoning in the park was an experienced ranger. Here again, sadly, we must refer to Lee Whittlesey's *Death in Yellowstone*. On April 11, 1927, the winter keeper at Old Faithful cleaned out the stream leading into the reservoir that provided the hotel with water, and brought some of the plants to the local ranger, Charles Phillips, to identify. He thought them to be camas—another potential misidentification—and he, the winter keeper, and the keeper's wife all ate some, Phillips eating two roots, and the winter keeper and his wife sharing one. That was all there was—happily. The winter keeper and his wife had a devilish time through that night and the next day, what with the nausea, vomiting, and convulsions; the wife was a "raving maniac," and the winter keeper not much better. The next day, April 12, they wondered where Phillips was; it was his habit to come every day at 8:00 p.m., to listen to the radio and visit—Old Faithful was a lonely place in that era, and it was still effectively winter. The winter keeper made his way through the snow to the ranger station, where he made a gruesome discovery:

> *He found the building dark and cold and Ranger Phillips dead on the kitchen floor. The ranger had apparently died sometime very early that morning.*
>
> *Phillips was lying on his back with one shoe unlaced and his clothing partly open. He apparently had had his bed slippers in his hand when the attack occurred, because they lay on the kitchen floor. Investigators found that he had apparently come out of the bedroom into the kitchen and fell forward, striking his head above his left eye on the table. Dirt on his hands indicated he had crawled on the floor for a while, probably in agony, and he exhibited the vomiting and frothing at the mouth which are characteristic of hemlock poisoning.*
>
> *Charles Phillips was widely read, educated, cheerful, and well liked by all in the park. He had a keen sense of humor and was an excellent writer and observer of the park's thermal features, having written pieces for the park's Ranger Naturalists' Manual. Had he not died, he probably would have written a book on Yellowstone's hot springs and geysers. A hot spring at Norris Geyser Basin was subsequently named for him as Phillips Caldron.*

That is a rare honor; most thermal features are named for a quality they exhibit, not a person. One lesson of this episode is that if it can happen to Phillips, it can happen to anyone.

It happened again in 1985. On August 3 of that year, Keith Marsh, a visitor from Nebraska, hiked with two companions to Heart Lake, in the backcountry south of Yellowstone Lake. With Mount Sheridan towering over its shore, it is one of the truly spectacular spots in the backcountry. Marsh found a plant and ate parts of it; one of his companions did the same, but spit it out because he did not like the taste. That is what taste is for. Though imperfect, it stops us from eating things we should not.

The plant was, of course, water hemlock. Whittlesey's description of the aftermath begins with yet *another* potential misidentification:

> *Marsh apparently mistook the plant for edible yampa root. During the evening, he went into violent convulsions. Rangers Brian O'Dea and Pat Ozment rode bicycles into Heart Lake just before midnight and found ranger Ann Marie Chytra with Marsh at Sheridan Creek. The rangers finally got an IV going, but it was too late for Marsh. He suffered at least seventeen seizures while rangers Chytra and Wes Miles gave him CPR and rescue breathing for many hours. "It was the most violent death I've seen," said Brian O'Dea later, acknowledging that it had taken Marsh eight painful hours to die.*

A third death from water hemlock poisoning may have occurred in 1971, when a woman named Virginia Hall arrived dead at the Lake hospital after sinking into convulsions. The autopsy was inconclusive, but she is supposed to have eaten "wild carrots" during her visit to the park.

So, stay away from this stuff, okay? "All natural" does not always mean "good."

The world of microscopic life is a world of superlatives, and of great beauty. It is also a world of horrors. Naturally, Yellowstone is well endowed in this area, as in all others.

Given the nature of that life, we do not notice it. The life I am talking about is the kind that eats you from the inside out, not the other way around.

And yes, we have lots of the dangerous kinds.

We should first get a few things straight. The kind of life I am talking about belongs to categories that are not interchangeable. Some are bacteria, single-celled organisms that might be a micrometer in width and a few micrometers in length. A micrometer is one one-millionth of a meter; the period at the end of this sentence is about five hundred micrometers wide. Humans love to consider themselves the dominant life-form on this planet, but bacteria really deserve that title. The oldest fossils are of a kind of bacteria, and they will survive long after humans are gone. They live everywhere, having been found in the outer reaches of the atmosphere, thirty miles up, and in the deepest trenches in the ocean. Yes, they can hurt us. Tetanus is caused by a bacterium ("bacteria" is plural). So are tuberculosis, pneumonia, cholera, botulism, and many others. It is bacteria, however, that antibiotics will kill, an innovation that largely dethroned bacteria as a killer of humans. They are also a key element in what scientists have come to call the human microbiome. As professors of the life sciences like to say, if every cell of your body were to suddenly disappear, you would not vanish. Instead, what would remain would be a ghostly human shape that would not immediately dissolve, but instead stand there giving everybody the creeps. It would mean them no harm. That specter is the microbiome, something in the vicinity of thirty-nine trillion microbes that live side-by-side with the thirty trillion cells that make up our actual "body," a thing that has an increasingly fuzzy definition. Most of these bacteria will not harm you, and without some of them you could not survive.

What antibiotics will absolutely not kill is viruses, in part because it has never been clear whether viruses are alive to begin with; writers have to fish for labels, calling it an "infectious agent" or "biological entity." A virus is just some DNA or RNA—genetic material—wrapped up in an envelope of protein. It lacks most of the qualities that define a living creature. It is inert; it does not metabolize; it does not have cells; it does not reproduce. At the same time, it is made of the same stuff we are, and so we might borrow a piece of internet slang to describe it: It is not alive, but it is "life-adjacent." Viruses replicate by hijacking a healthy cell and turning it unhealthy, or dead, in the process of reproducing. They are, of course, best known for making you sick. They cause the common cold; they cause Ebola hemorrhagic fever; they cause all kinds of diseases in between.

Other microbes we need to concern ourselves with are archaea and protists. Archaea get their name from the Greek word for "ancient," which gives us *archaic* and *archaeology*. They got this name because they separated from all other living things billions of years ago. They do unique things, like producing methane. They also live in the most extreme environments. If a place is too hot, cold, salty, alkaline, acidic, or radioactive for anything else to live there, chances are good some kind of archaea is thriving there. Yellowstone might have been set aside as a preserve for archaea, so many different kinds live here. Finally, when we talk about Yellowstone microbes, we will also be discussing protists, microscopic creatures that, unlike the others I have listed, have a nucleus in their single cell. If you are old enough, as I am, you might remember the biology teacher in junior high calling "Protista" a kingdom of life, a division that scientists no longer recognize, although they still call them protists. Yellowstone might be a preserve for them, too, and they also live in extreme environments. The group includes amoebae, algae, and protozoans, and as is so often the case, the examples we find in Yellowstone are often spectacular. The algae that lights up the ground around the thermal features are protists, gaudy almost—if not quite—to a fault.

Yellowstone has contributed much to the understanding scientists have of the weird world of the microbe. Starting in the mid-1960s, Thomas Brock, a microbiologist with Indiana University and later the University of Wisconsin–Madison, began studying Yellowstone's hot springs to see what might be living there. The thought at the time was that anything living in the very hottest of the park's water would be, in effect, barely hanging on; it would be doing well to merely survive in an environment that would always be on the verge of killing it. Brock, his wife Kathie, and a long line of student assistants found otherwise. Working under primitive conditions, in tourist cabins and dilapidated bunkhouses, they discovered a whole universe of organisms living in water that should, according to the conventional wisdom, have cooked them to death. They were named hyperthermophiles, a new classification for life-forms that live in water above 140°F, and often above 176°F.

They do not merely survive there; they thrive there, indeed cannot live anywhere else. The most extreme of these extremophiles (another name for them) have been found around hydrothermal vents on the floor of the ocean, where they spend their lives in water above 250°F—and remember, at sea level, water boils at 212°F. And some of the life-forms in Yellowstone are actually *polyextremophiles*. To the ancient Greek word-ending *phile*, meaning "one who loves," is added the equally Greek prefix *poly*, meaning "many." Some of our Yellowstone microbes can handle more than one kind of extremity. The hot springs are hot, of course, but are also blisteringly acidic. The polyextremophiles take that in stride. They have taught us much about what life was like at its very beginning, when the earth was one big extreme environment. They have also shown us what life might be like on other planets, holding out hope that we might find life on Jupiter's moon Europa, home to both volcanoes and water.

Because they are adapted to these environments, the polyextremophiles pose no threat to us. They would find us insipid, and altogether too cold. These life-forms do not just survive in hot, acidic water; they require it. The inside of a human, at a mere 98.6°F, is not anywhere near warm enough to get them going. The germs that hurt us live at that temperature, and that is why boiling water, or the hospital's autoclave, kills them.

But places in Greater Yellowstone can function as refuges that are closer to that cozy 98.6°F, where other microbes can live and either quietly hide, or—less often—run riot and kill. They can do us a great deal of harm.

Some years ago, for instance, researchers looking at the mountain lions in the park got an unpleasant surprise. We will learn more about mountain lions later in this book, but what we are interested in here is what turned up inside them. It was about the most alarming kind of bacteria you can imagine: They tested positive for plague. The researchers were not shocked; these things happen in the world the mountain lion lives in, although it was one more piece of bad news for a population already under stress.

It may be that most people are inclined to imagine that plague (bubonic plague is one variety) no longer even exists. It very much does. According to the CDC, between the years 2000 and 2022, an average of seven people per year developed plague. Because we have antibiotics now, it is no longer a death sentence, although it certainly can kill; as a popular physician's diagnostic handbook, *The Merck Manual for Healthcare Professionals*, puts it, "Before antibiotics (1900–1941), case fatality among those infected with plague in the United States was 66%. By 1990–2010, antibiotic treatment of plague reduced case fatality to 11%"—still surprisingly high, given that the microbe to blame is a plain old bacterium, and not a virus. Antibiotics will kill it before it can kill you (Merck recommends streptomycin and gentamicin, which some of you might have as leftovers in your medicine cabinet). Some years, no one dies of it; other years, one or two of its victims, and sometimes three, get to the hospital when it is already too late, or the doctors do not realize until too late what they are dealing with, and they *do* die. The doctor's challenge is to identify it before it does too much harm, but as with so many diseases, plague begins with fever, chills, and headache. It is what doctors call a zebra, after the saying they learn in medical school: "When you hear hoofbeats in Texas, look for horses, not zebras." When confronted with a patient who has fever, chills, and a headache, it is usually a waste of time to fret about plague. Except sometimes . . .

Another fact about plague that most people do not know is that the big outbreaks in history are not separated from us by great stretches of time. Some, to be sure, are. Historians recognize three pandemics. The First Pandemic occurred in the waning years of the Roman Empire, technically the Byzantine Empire, and is still sometimes called the Plague of Justinian, after the emperor reigning at the time, who came down with plague and survived it. It began in Egypt in the year 541, and reached the capital, Constantinople, by the next year. At its height, it was killing perhaps five thousand people a day in the capital, and when it was through, it had carried off about a quarter of the population of the Eastern Mediterranean; estimates of the actual number bounce between twenty-five and one hundred million dead. Reaching a reliable number for ancient epidemics is more or less impossible. The plague became endemic in the region, and did not disappear entirely until perhaps the year 750.

It was just getting started. There are in the world such things as plague "foci," the plural of "focus." We have some in the United States. They are places where plague exists all the time, in burrowing animals; the plague pathogen survives in the soil as well as in the animals, and so is able to get through the winter and become a permanent resident even in cold places like Greater Yellowstone. In California, for instance, is the San Bruno Mountain plague focus, which one historian (Charles Gregg, in his excellent 1978 book *Plague: An Ancient Disease in the Twentieth Century*) calls the "most intensively studied plague focus in America." One species of vole living there, the California vole, is immune to plague, although their bodies are full of it. Harvest mice in the area had plague go through their community in the spring of 1963, and were "virtually exterminated." What happens is that a group of burrowing rodents will leave a plague focus and hitch a ride in human conveyances, in ships and baggage, and so descend upon a human community that has not known plague and so has no immune defenses against it. That is what happened in the plague outbreak you likely have heard about, the Black Death of the Middle Ages. According to one scenario, plague is thought to have jumped out of populations of marmots in Central Asia and into rats; it was then carried along the Silk Road, the medieval trade route that linked China and the Far East with Europe. It arrived in the Crimea first, at the trading port of Kaffa, owned and operated by the city-state of Genoa; it was carried by sea around the Mediterranean, and by 1347 had a solid foothold in, among other places, southern Europe. The rest, as they say, is history. Medieval writers thought that a third of all Europeans died. Modern estimates make the death toll maybe as much as half.

It did not stop. The period between 1346 and 1353 was a high point, or low point, for the disease, but after the latter date, people kept dying. The Second Pandemic lasted through the eighteenth century and into the nineteenth. William Shakespeare, in the sixteenth and seventeenth centuries, wrote about plague all the time. When he had Mercutio say, in

Romeo and Juliet, "A plague o' both your houses!" the audience knew just what he was talking about.

It came back in the 1890s, with the first deaths happening decades earlier, and by some accounts, this Third Pandemic lasted until 1960. That the plague was now happening in the modern era seemed in places not to matter: Countries like India and China still ran up medieval death tolls, with millions of people perishing miserably. But it was now, also, that we learned what we now know about the disease, and so at last vanquished it as a major killer. In 1894, Alexandre Yersin, a Swiss-born bacteriologist who had trained in Louis Pasteur's research laboratory, went to Hong Kong to try to get to the bottom of the plague outbreak there. He succeeded: He found the bacterium responsible, now named *Yersinia pestis* in his honor (*pestis* is Latin for "plague"). During this period, the route of transmission was also worked out, after Yersin found his pathogen in rodents, too. Scientists then drew the famous triangle of rat, flea, and human victim, in which the fleas transfer the disease from rat to rat, and when the rats die of plague, the fleas jump onto and drink from a new victim: humans. Yersin did his work under primitive and highly dangerous conditions, and in competition with much better-funded scientists. People have a soft spot for Yersin that shows up in otherwise coldly formal histories. We all love a scientific underdog, same as we love underdogs of all kinds.

Researchers later came to understand the behavior of *Yersinia pestis* in the human body. "Plague" is a general term (and although it looks like some kind of grim joke, that is the actual name of the disease, the name it goes by in medical textbooks). The bacteria may get into the lymphatic system; it will then end up in the lymph nodes nearest the flea bite. The lymph nodes bulge and protrude and literally rot, swelling and turning black. These are called "buboes," and give bubonic plague its name. Sometimes the *Y. pestis* gets into the blood, multiplying there and, among various other horrors, causing disseminated intravascular coagulation. For this one, you might ask for clarification not from your doctor, but your plumber. It is clotting that happens inside blood vessels and in too many places, causing tissue to die from lack of circulation. The flesh turns a dark hue tending toward black—hence the name, the Black Death. Worst of all, perhaps, is pneumonic plague, which can develop out of the other two varieties, or can come out of thin air, literally. This form of plague occurs when the bacteria find their way to the lungs; when that happens, the victim will be breathing out aerosol particles that are little different from the stuff that issues from bioweapons. The plague can spread rapidly when this happens, especially in situations in which a great many people are living packed together in impoverished circumstances. The medieval Black Death is thought to have been significantly pneumonic. The death rate from this variety is high. The last major plague outbreak in the United States, and also the last time pneumonic plague is known to have developed here, occurred in Los Angeles in 1924 among Mexican migrants who lived packed together in substandard housing. The first victim probably got the disease from a dead rat, but from there, plague was passed around by aerosol transmission, and the bodies accumulated. At least thirty-seven people developed plague and were clinically diagnosed. Of these, thirty died.

We live now amid aftershocks of the Third Pandemic, although I should note that plague is not as well understood as it might be if it were more common today, and therefore more heavily studied. It is only recently, for instance, that *Yersinia pestis* has been proven to be the agent responsible for the three pandemics. Scientists can now find bacteria in the teeth of people who died of plague and amplify its DNA, and as a consequence, *Y. pestis* is known to have been killing people for thousands of years. In the great mass graves filled with victims of the Black Death, it turns up routinely. Still unclear is how this species of bacteria got to the New World. For decades, it was assumed that it came in a ship during the Third Pandemic, specifically in one of the ships suspected of bringing plague to San Francisco in 1900, leading to an outbreak in that city that killed at least 119 people. There was also an outbreak in Seattle. Another possible explanation, however, is that it came in animals across Beringia, the now-submerged land corridor between Alaska and the Russian Far East that was high and relatively dry during the ice ages. Charles Gregg explains, noting first that the great eruptions of plague earlier in history can be traced back to various Old World species of marmot:

The marmot is a native of North America that migrated across the Bering land bridge into northern Asia. The native North Americans came the other way . . . [and] plague may have come with them. According to this view, plague was entrenched in the wild animals of North America centuries ago; it was simply recognized for the first time during the early years of this century following the outbreaks in San Francisco and Seattle. There are some distinctly inhospitable areas between San Francisco and western Kansas [the farthest reach eastward of plague during the Third Pandemic]. . . . Some scholars question whether wily Y. pestis, *for all its low cunning, could have made the journey in such a short time by the slow transfer from one rodent host to the next most eastward one, in a sort of pestilential bucket brigade.*

Instead, it was there long before Columbus's voyages. If so, then here is an interesting possibility: *Yersinia pestis* may not be a nonnative species, but a native. Theoretically, the government would then be required to protect it, especially given its obvious historical significance.

Yellowstone has marmots. You can see them at, among other places, the Upper Geyser Basin, where they occupy the ridge behind Grand Geyser, just downstream from Old Faithful. Watch for the brown animals that look like small beavers, or very large squirrels, ambling around and loudly *peep*-ing at each other; these are yellow-bellied marmots, a kind of ground squirrel, actually. A marmot is not going to give you plague, and neither is a mountain lion. Presumably, a lion would give you the illness in the traditional

A Yellowstone marmot, in a pile of volcanic rock near Undine Falls.

Another Yellowstone marmot, this one using its adorability to panhandle near Bridge Bay.

way, by letting its fleas bite you—and if you are getting intimate enough with a mountain lion to share its fleas, you have a more immediate problem than mere bubonic plague.

Nevertheless, plague is resident in Greater Yellowstone. It was biologists working for an organization called the Teton Cougar Project who found *Yersinia pestis* in Greater Yellowstone mountain lions (the animal goes by a variety of names—mountain lion, cougar, panther, puma—but is still the same animal). The scientists examined lions from Jackson Hole, the spectacular valley that is partly enclosed within Grand Teton National Park, the southern sister to Yellowstone National Park. Their findings were published in 2020 in the scientific journal *Environmental Conservation*, in an article authored by Howard Quigley, Mark Elbroch, and veterinarian Winston Vickers. *Smithsonian* magazine summarized the project and its findings: "Over nearly a decade, between 2005 and 2014, the researchers checked 28 cougars for signs of *Y. pestis*. Eleven of the cats were found after they'd died, and four of those . . . died of the plague. The researchers drew blood from 17 other cougars and analyzed the samples for antibodies, the chemical footprint left behind after the immune system fights off an infection. Eight of the 17 tests came back positive. In all, about 43 percent of the cougars studied faced *Y. pestis* infections."

Plague has claimed human victims in Greater Yellowstone. Between July 26 and August 3, 2008—the precise height of the tourist season, by the way—a Connecticut teenager visited the area to do service projects with the Boy Scouts, and he came away with a souvenir. Once home, he landed in the hospital, with fever, malaise, and a prominent swelling on the side of his neck: the classic bubo of bubonic plague. The doctors hit it with antibiotics, and he made a full recovery. The CDC investigated, along with the state health authorities. The Scout had stayed in various places around Jackson Hole and Yellowstone National Park, including the campground at Grant Village, on Yellowstone Lake. "Not surprisingly, the exact location where the young man was exposed to the disease could not be determined," the Wyoming state epidemiologist said in October of that year, having predicted that outcome back in August. Any mammal can harbor plague, and all it took was one flea bite.

And it is at this point in a piece of writing of this sort that I am supposed to give you a breathless warning about the great peril you are in. You are not, and I am not going to write that warning. Do not feed the animals, even the little ones like chipmunks. Do not get so intimate with rodents that their fleas are biting you. Do not get intimate with the mountain lions at all. You will be fine—the *Yersinia pestis* is mostly tucked away in rodent burrows (it has to be, to survive the winter), and avoiding it is a matter of just being more or less normally human. We can only come into contact with it under extraordinary conditions.

Besides, Yosemite has had a much worse time of it with plague than Yellowstone has, and the specifics of those cases show why worry is pointless. In 1959, another Boy Scout camped in the park and came down with plague. The incident is almost forgotten today, but you can read about it if you have the patience to dig—it is online—into a charmingly primitive publication called *California Vector Views*, the in-house newsletter of the California Bureau of Vector Control, typewritten and run off on a mimeograph, or maybe a Ditto machine, hard to say. After the Scout fell sick, on June 23, 1959, investigators from the US Public Health Service retraced the route his party had followed along Yosemite Creek, north of the famous valley, and identified their campsites. This work requires a certain level of courage, as is clear from the report on the investigation in *Vector Views*: "A second campsite they had used was pinpointed. In this area, and for a radius of several hundred yards, ground squirrels and chipmunks had been severely decimated." There had been a Black Death among the rodents, and the kids had camped right in the middle of this field of skulls. The Public Health Service investigators only recognized the place for the horror it was because of their specialized training. Unless you have such training or are an epidemiologist with the CDC, you cannot recognize the danger anyway, on the rare occasions when it occurs. You would have camped there, too.

Worse was 2015, when *two* plague cases occurred after exposure in Yosemite, both diagnosed in August following earlier visits. Impossible, surely . . . except when we remember what happened in Glacier National Park in 1967 in a single night. A fourteen-year-old boy in Los Angeles developed septicemic plague, and an eighteen-year-old young lady in Georgia developed the bubonic variety, both after spending a significant amount of time in Yosemite. Both recovered. The California Department of Public Health, the Centers for Disease Control, and the National Park Service investigated with a vengeance and found that the fourteen-year-old had duplicated the experience of the Boy Scout in 1959; in the place where he camped, plague had nuked the rodent population. The eighteen-year-old had visited the famous Glacier Point, where the rodents were both friendly to people and positive for plague. There was plague elsewhere, revealed by an expanded investigation. The news media reacted as we might expect. *Slate* magazine, reaching for the stars, managed to blame the cases on racism (their headline: "Why the United States Is Plagued With Plague: The history of a racist, bungled public health battle").

It is a truth universally acknowledged that in such a situation, the government must be to blame. We should resist scratching that itch. Yosemite is an enormous park (and Yellowstone is bigger). The animals involved are tiny. And all it takes, as we have seen, is one bite. By a flea.

Besides, there are so many other pathogens to worry about. And the helpers that give them a ride, too.

A "disease vector" is any creature that can pick up a pathogen, move it around, and transmit it to another living thing. When the subject comes up, the kind of creature usually being discussed is an arthropod, a member of the phylum Arthropoda, which, for our purposes, is mostly what we think of as bugs: arachnids, like ticks and spiders, and insects (which group, I should note, is gigantic; the arthropods that will come up here are relatives of lobsters and horseshoe crabs, and not even all that distant). At our point in history, when biologists put living beings into groups, they are usually talking about genetic relationships, but the defining characteristic that makes a vector a vector is usually the nice, down-to-earth way they have of making a living: They drink blood. Vectors are bugs that feed on your living essence: lice, fleas, ticks, mites, flies, sand flies, and mosquitoes.

We have already looked at one of the classic instances of a disease outbreak in which vectors played a major role: The great eruptions of plague through history were driven by vector-borne transmission. Famously—in, say, the overstuffed cities of Europe during the fourteenth century—the rats had the plague first. When a rat died of the disease, its fleas departed and, with the rat population gone or going fast, they needed a new host. Fleas are adapted to specific victims; the two evolve together, and the fleas would rather not move—but sometimes the alternative is death. It is thought that this movement results from an adaptation by *Yersinia pestis*. The pathogen "wants" to spread to new victims, and killing its host forces the fleas to get up and move. So it was that millions of Europeans were bitten by insects that were full of plague. It is not the only way to get the plague—you can get it merely by handling an animal that has died of it—but it is likely the most famous host-vector-disease relationship in history.

Yellowstone is not especially flea-ridden, but another of history's great insect vectors is available in the region in really appalling numbers. If you are presently in Yellowstone National Park, especially in the spring, you will have noticed them, their presence in some places a good deal more oppressive than in others: mosquitoes, squadrons of them—no, whole air forces of them, formed up in attack posture. In places, they get so thick in the spring that hikers will repeatedly suck individual insects into their open mouths. When that happens, the creature ends up so far back in the throat that the only option is to swallow it. Parts of the park are locally infamous for the intensity of the visitation; among employees, for instance, Grant Village has traditionally been known as The Mosquito Coast. There are no fewer than twenty-nine species of mosquito in Yellowstone National Park; another source says there are thirty-three.

There is so much diversity in their local populations that they even have an infectious agent of their own: You may be pleased to learn that, in Grebe Lake and DeLacy Creek, local populations of mosquito larvae are vexed by their own species of *very* small nematode parasites. Mosquitoes are, of course, the vector that spreads malaria, which was once common in the United States. It was only eradicated here in 1951, and can still pop up; after a twenty-year run with none at all, a number of locally acquired malaria cases occurred in 2023, in various parts of the United States. Rest easy, though; at no point in its history did Yellowstone have malaria, or the other big mosquito-borne diseases, yellow fever and dengue.

But we do have ticks, although when one asks around, the level of disagreement about how many can be amusing. There is (of course) a subreddit on Reddit called "r/yellowstone." In it, an innocent prospective visitor once asked about spending a few nights in an RV park near West Yellowstone with his dog. Would the dog pick up ticks? The "discussion" devolved into almost pure hysteria: The dog would be eaten alive by thousands of ticks, and dispatching the poor animal with a bullet behind the ear would be the only humane treatment. If you know Reddit, you know the kind of thing I am talking about (visiting the site, I wondered if they would warn against bears attacking drivers on Interstate 90). After close to fifty increasingly frenzied responses, a sane person intervened. Yes, there are ticks, especially in the spring, and especially in a wet spring. They are concentrated most heavily at the lowest Yellowstone elevations, between 4,000 and 6,500 feet. They can appear as early as February at the lowest elevations, but seem to peak in May, as long as the weather trends wet; if it dries adequately, they may skip that year, mostly.

Scientists have a fun name for tick feeding behavior: Ticks cannot fly, so in a process called "questing," they hang from a piece of vegetation and release when a victim brushes past. A trail with no vegetation along the sides has little risk. When hiking in the early spring, locals in Gardiner and the Yellowstone River valley often wear gaiters (lower leg protection, like the leggings or "puttees" worn by soldiers during the world wars), and they do so expressly to repel ticks. Dog owners report that their pets pick up ticks, although they seem to mostly do without the kind of medicated collars and chewables that dogs need elsewhere. (In fairness, Montana dogs do without most luxuries.) Inside the national parks, the authorities require that dogs stay on a leash at all times; it keeps them away from the bears, and it keeps them from jumping into water that is close to boiling. Outside the park, dogs go where they wish (although in some places in national forests, like campgrounds, leashes are mandatory). A key skill possessed by veterinarians in Montana is the removal of porcupine quills from dog noses. They do it all the time.

    I must confess that I find it all baffling. During the last thirty-five years, I have hiked thousands of miles in Greater Yellowstone, many thousands, most of it on the northern range, inside the exact altitude range in which ticks are supposed to thrive. I have looked for ticks, actually consciously looked for them. Ticks are relatively easy to identify because they are arachnids, with eight legs instead of the insect's six—and if you see a creature you think may be a tick, count the legs, and then, if you have the stomach for it, take a photograph and enlarge the image. There are species of spider in Yellowstone that look a great deal like ticks, and are harmless by comparison. They are easy to tell apart: Spiders have a "waist," a middle section called a cephalothorax, while a tick is all abdomen. It looks like an apple seed come to life.

    So I know what to look for—yet in all this time, I have seen exactly two ticks, neither of them actually on me. It is true that ticks do seem to favor some people over others. The same kind of mythology that comes up with mosquitoes appears here, too: A common belief is that ticks like people with diabetes, and people wearing perfume, but neither is true. There is some evidence that they favor type A blood over other types. The bias is not overwhelming enough to explain my good luck with ticks—and anyway, I *have* type A blood.

    I remember the number—two ticks—partly because the occasions were memorable. Once, the tick was attached to the impossibly perfect, impossibly vulnerable skin of my infant son's neck. The other happened many years earlier, when I found a tick on the floor of the bunkhouse we lived in when I worked at Old Faithful. I immediately called a friend, the first of our circle of friends to have landed what, in the language of national park employees, is called a "real job." Without saying hello, I announced, with some enthusiasm, "I found a tick!" The real job she had gotten was in the laboratory of a renowned expert on tick-borne diseases at the National Institutes of Health laboratory in Hamilton, Montana.

In a less enthusiastic tone, I noted that I had heard Lyme disease was not that big a thing around here. Somewhere or other, I had picked up the—correct—information that tick-borne diseases are not the same kind of problem here as in, say, the wet, humid Northeast.

"No," she said, laughing. "You'll just get Rocky Mountain spotted fever and die."

I was absurdly young, and somewhat more clueless than I am now, so I had no idea what she was talking about. If states had, along with the state flower and state bird, a state disease, Rocky Mountain spotted fever could be Montana's, and its causative agent could be its state pathogen. If a disease can be a force for good, this one was, because it did much for our early understanding of vector-borne diseases. During the nineteenth century, it was almost common in the Bitterroot River valley of western Montana, where the National Institutes of Health today does research on it, in that lab where my friend got her first-in-our-circle "real job" (last I heard, they were training her to work in the Biosafety Level 4 facility there, the kind of facility that they make bad movies about, where researchers in space suits work with the deadliest pathogens known to humanity. When we heard about that, we stopped being jealous).

Here, and elsewhere in Montana and Idaho, the crucial early breakthroughs were made. Rocky Mountain spotted fever is a nasty bacterial disease that starts with a fever and develops into a rash composed of what doctors call purpura, from the Latin word for "purple," bleeding of that color underneath the skin that makes for an unhealthy freckling and gives the disease the "spotted" part of its name. The freckles are not the most serious part of it, although they are related to the most serious part. The disease affects the lining of the blood vessels, and so can wreak havoc all over the body. An extreme result can be gangrene, leading to the destruction and amputation of fingers, toes, and whole limbs.

Rocky Mountain spotted fever is caused by *Rickettsia rickettsii*, and the Ricketts for whom it is named deserves the honor; Howard Ricketts died young in Mexico studying typhus, another disease carried by a vector and caused by a *Rickettsia* bacterium. It was probably inevitable,

Ricketts had so intimate a relationship with bacteria. He was known to "study" diseases by deliberately catching them. The intimate relationship continues, as whole swaths of the bacterial world carry his name.

Because it is bacterial, antibiotics will kill it. You get Rocky Mountain spotted fever from ticks, of course, from multiple species. The one we need to worry about most in Yellowstone would be the number one contender for Montana's state tick, the Rocky Mountain wood tick, *Dermacentor andersoni*. It is another of those life-forms that is native to Yellowstone, and that—given a really rigorous interpretation of the law—the National Park Service should logically be pledged to protect, along with *Rickettsia rickettsii*.

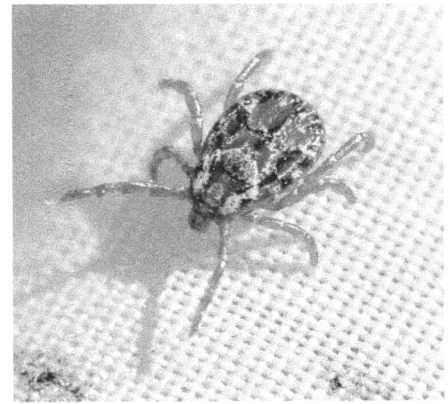

One of our Rocky Mountain wood ticks.
NATIONAL PARK SERVICE

But not Lyme disease. Perhaps the most famous tick-borne illness, Lyme disease causes a wide range of symptoms including, most prominently, fatigue, headache, and fever (some of the early cases were mistaken for juvenile rheumatoid arthritis). It is named for the towns, Lyme and Old Lyme, in Connecticut that were among the places where the disease was first recognized, and Connecticut is where you want to go if, in emulation of Howard Ricketts, you want to catch it. A risk map of North America will paint the Northeast black, while the Northern Rockies are bright white, appearing to have no risk at all; Lyme disease is the most common tick-borne illness in Montana, but all the cases originate elsewhere, because the tick that plays the villain does not live here. You get Lyme disease from deer ticks, also called black-legged ticks, *Ixodes scapularis*; the pathogen, in North America, is usually *Borrelia burgdorferi*, named for the scientist who isolated it in the 1980s, Willy Burgdorfer. Our friend, the one who had the first real job—Willy Burgdorfer was her boss. It really was a remarkable job to get as a first real job. Of course, the pay sucked.

Keeping track of tick-borne diseases in Yellowstone National Park is complicated by the fact that the National Park Service is not here to do epidemiology. In fact, they are no good at it at all. Happily, the Centers for Disease Control is here, reporting on affairs in Idaho, Wyoming, and Montana. There are five counties that either surround or share terrain with Yellowstone National Park: Fremont in Idaho, Gallatin and Park in Montana, and Teton and a second Park in Wyoming. In these counties, the CDC tracks a number of tick-borne diseases: anaplasmosis, babesiosis, ehrlichiosis, Lyme disease, tularemia, and spotted fever rickettsiosis, and how is that for a catalog of nightmares? The names alone make you ill.

These disorders are rare: In these five counties, Lyme disease is the clear winner in frequency (seven cases in Gallatin County between 2019 and 2022, two in Park County, Montana, and one in Teton County, Wyoming), and it is not even from here; people get the disease visiting some other part of the country, or world, and then see the doctor when they get home. In fact, the diseases are so rare that we can really not worry about them, except for one, and the creature that causes it qualifies as another Yellowstone native life-form: tularemia. It is not especially common, but by the middle of 2024, there had been eight cases in Wyoming where there would normally be two, enough that the state health authorities announced the fact in the news media and asked people to be vigilant. There was a case in Teton County recently—a case in a human, that is—and a dead marmot floating in a lake near Jackson, Wyoming, tested positive for tularemia in 2018. The lake was Crater Lake (not the famous one in Oregon—a little Crater Lake), and a worry in that case was hikers drinking water from the lake. You can catch the disease by an amazingly wide variety of routes, and one is through water contaminated with the pathogen.

That pathogen is a bacterium, *Francisella tularensis*. It turned up in some ground squirrels in Tulare County, California, in 1911, and that is how the disease, tularemia, got its name. This bacterium, which moves around through so many different routes, also turns up in an amazing range of animals. Tularemia is still sometimes called rabbit fever, and important hosts include deer, a variety of rodents, and "galliform" birds—that is, ground-feeding birds like chickens—but it also turns up in all sorts of mammals, amphibians, birds, reptiles, and even fish. You can get it by handling or eating infected animals (it has always been associated with hunters), but also from insect bites, and the bacterium is so easily aerosolized—that is, turned into fine particles that will then blow away—that it can be caught literally out of thin air. There was an outbreak on Martha's Vinyard in 2000 that was linked to lawn mowers; mowing your lawn or cutting your brush is one route by which the disease can be caught. It is intensely pathogenic—that is, it causes disease easily, so easily that you can get it by taking in as few as ten to fifty individual bacteria. It is so infective that scientists have to be especially careful, and laboratory exposure is a significant threat. This infectiousness, and the fact that it is so easily aerosolized, means that it has long been of interest to governments developing biological weapons. Japan experimented with it during the Second World War, and during the Cold War, both the United States and the Soviet Union maintained stockpiles of it.

It can be fatal, but *Francisella tularensis* is a bacterium. Antibiotics will kill it. It begins, as so often seems to be the case, with manifestations that look just like the flu, then causes a range of symptoms that depend on how the victim gets it originally (for example, breathing it in while mowing the lawn leads to a "pneumonic" tularemia). The one common symptom, however, is a marked fever, running up to 104 degrees. The insect vectors are deerflies, horseflies, possibly mosquitoes, and definitely ticks.

The first and last word on this general subject is an article that ran in *Yellowstone Science* back in 1996: "Yellowstone's Insect Vampires," by John F. Burger. It has never been surpassed. Burger is Yellowstone's guru of arthropod vectors, an emeritus professor at the University of New Hampshire who, like so many of us, got hooked on the place working here as a kid, in his case for the now-extinct Yellowstone Park Company, as a dishwasher at Fishing Bridge in 1959—a "dish dog" in the slang of the time, I have been told. He possesses the kind of bag of tricks that develops in a scientist who spends decades here, like his unique manner of spotting a swarm of biting flies from far away, even through binoculars:

"Deer are just tormented by these things," he told an interviewer from *Yellowstone Science* in 2007. "We have a tail-flicking index. You can tell the severity of biting flies by the number of ear flicks or tail switches per minute. You can drive by an area and, depending on what the tails are doing, you can determine what the situation is out there. At Mammoth, in the evening, sometimes the ears on the elk are just going a mile a minute. That's a good index for activity." There exists, by the way—and I'm not surprised that Burger knows this—one species of mosquito in Yellowstone that feeds not on elk, or deer, or bison, or humans—but on frogs, and nothing else.

I'm not surprised, because Burger's knowledge is pretty close to encyclopedic, and he has searched for, and found, insects in the most improbable corners of the park. The interviewer from *Yellowstone Science* asked him, among other things, about vector-borne diseases. Burger covered the same sort of issues we have here, but added an odd one: "A species that's a vector of West Nile Virus occurs in the park, *Culex tarsalis*." The common name is the western encephalitis mosquito; as the name suggests, it also carries St. Louis encephalitis, which happily does not occur in Yellowstone. "But I'm not certain," he continued, "how abundant it is in the park. I did find it in Frying Pan Spring one year. Some of the pools there aren't as hot as some of the others, but I remember thinking, mosquito larvae in Frying Pan Spring? I mean, whoa! But that's what it turned out to be—*Culex tarsalis*." Frying Pan Spring is right next to the road, complete with a parking area, just north of Norris Geyser Basin. You can park there and ask yourself how likely it is that mosquitoes would turn up in that water, but there they are. You should not worry too much about West Nile virus, which is not a native, and is rare enough that we could skip it—and its close relative Zika virus (yellow fever virus and dengue virus belong to the same genus, and all these are really creatures of the tropical world). When he talks about insects, however, Burger regularly notes how much we do *not* know. His science, entomology, is still at a remarkably early, even primitive level of development. As he said in the interview:

When you come to Yellowstone you think elk, bison, bears, wolves. This happens very frequently—often when I'm coming back from hiking in Pelican Valley, people are asking, "did you see any animals?" I say, "lots of animals." "Oh, what did you see?" "Well, butterflies, bees, flies." "Oh. Okay." And they walk off. I think most people who come here just visiting have heard about the charismatic megafauna, and that's what almost all people focus on. But if you look at ecosystems, you start thinking about how many insects there are in Yellowstone relative to other animals. We probably don't know even within an order of magnitude how many there are. If you were to ask me for a number, I would say many thousands, but that would just be a guess.

Looking out across a meadow, ask yourself: Which animals consume the greatest mass grazing, the elk and bison, or the grasshoppers? At this point in history, no one knows.

Burger is the kind of Yellowstone researcher who is an especially useful person to have around, because he knows the answers to the questions we all wonder about, but can never find a satisfactory answer to (and no, the internet is not much help, because it is so loaded with rubbish). What attracts mosquitoes? "They are attracted by movement, CO_2, heat, moisture, and volatile compounds on a host animal's skin surface," he writes, in "Yellowstone's Insect Vampires." The phrase "volatile compounds" sounds odd to the rest of us, as if he were referring to explosives, but Burger is a scientist; a volatile substance is one that evaporates readily. He is referring here, yes, to perfumes, and also lotions, colognes, and other stuff we put on our skin. If they were not volatile, we could not smell them. They are bleeding molecules all the time, leaving a trail that the mosquitoes follow right to your skin.

Burger's specialty is actually biting flies. It is likely that nearly every visitor comes to Yellowstone knowing that the mosquitoes will be a problem, at least in June and July, and in some places longer. Our flies, however, are often a nasty surprise.

We have horn flies, stable flies, and moose flies, all belonging to the same family, and all, happily, a nuisance mainly to large animals. We have biting midges, which you may know as "punkies" or "no-see-ums."

Scientists do not know much about the local species, and some of what they do know, you may have learned already, the hard way (they cannot bite through cloth, but somehow have no problem passing through a window screen, and their "painful bites are wildly disproportional to their size"). We have buffalo gnats, named for the hump on their backs. Here, the happy news is that the populations are highly localized; they like big rivers like the Lamar and Yellowstone, but they also like obscure springs and creeks that you are not likely to visit. Here, as with all of these creatures, it is the animals that really suffer (indeed, in horrifying ways: "Running a hand over the chest of a horse can cause a hundred or more feeding buffalo gnats to drop out of the hair"). We also have a beast you will not have heard of, snipe flies, which are often confused for other species, but are uniquely nasty in their own way. With this one, the happy news is that, for reasons that remain obscure, they really only make themselves a great nuisance in certain years. Yellowstone people will remember 1967, 1994, and 2006 as banner years. As Burger puts it, "1994 was the year of the snipe fly."

(When they see lines like that, city slickers may be tempted to chuckle at our expense: "Their high point was flies!" they think. But when you are fighting snipe flies in the middle of a snipe fly year, all other forms of human endeavor shrink to insignificance.)

The biting flies that you should most concern yourself with are deerflies and horseflies. It's the classic bind: You may not be interested in them, but they are intensely, painfully interested in you.

In certain places and during certain years, the deerflies and horseflies are overwhelming. Deerflies are about the size of a housefly, with transparent wings that have a distinctive dark crossband; horseflies look rather like honeybees, except that they have only a single pair of wings. Both species like to bite, and in a bad year, the horseflies are like science fiction. They are almost unbelievably robust, surviving a surprisingly long time after having been ripped in two. The old Montana cowboy joke comes up regularly, the one about how if you slap a horsefly on your horse's neck, the horse dies of a broken neck, and the fly is not even inconvenienced.

Burger gives us the facts we need to know about them:

> *Horse flies and deer flies are most active on warm, sunny days between 10 a.m. and 4 p.m. and less active on cloudy or cool days or during rainy periods. These flies are strongly visually oriented, and usually depend on host movement for long-distance attraction, and on host odors, color, and CO_2 for close attraction. Their blade-like mouthparts for piercing and cutting through skin can produce very painful bites. Deer flies tend to be most common in meadows or along trails, where they perch on vegetation and fly out and around objects moving on the trail. They usually circle once or twice while determining whether to land and attempt to feed. Horse flies also rest in vegetation and fly out after moving objects, particularly in open areas. Trails and large meadows are prime locations for* Chrysops fulvaster, *a vicious biting deer fly that is extremely abundant in large meadows and near some thermal areas.*

I may have had improbable luck with ticks, but *Chrysops fulvaster* has ripped me to shreds. They *adore* me.

But getting back to ticks (and Burger, naturally, is aware that they are arachnids, not insects), Burger gives us the lowdown. There are two main species in the park, the winter tick, and the one we have met already, the Rocky Mountain wood tick ("although other species likely appear in the park," and note the "likely"—it is a big place, these are small creatures, and entomology as a field does not have enough money or personnel to find out the truth). As humans, we do not have to worry about winter ticks; they target other animals, including the big grazing animals you are likely to see in the park, and in such numbers that individual animals may "carry hundreds or even thousands of ticks," so many that such animals "are almost invariably in poor condition and are likely to die during the winter" (Ernest Thompson Seton, one of the founders of the Boy Scouts and a big fan of Yellowstone, said that moose were victimized by winter ticks worse than by "wolves, bears, and cougars").

It is those Rocky Mountain wood ticks that humans have to worry about, and aspects of their lives are strange. They go through phases, kind of like adolescents, or the creatures in the *Alien* movies, you be the judge. Burger compares it to the winter tick, a "one-host" species:

> *As a "three-host" tick, the Rocky Mountain wood tick has a very different life cycle. After hatching, the larva feeds on a small mammal such as a ground squirrel, then drops off and molts to the nymph stage. The nymph feeds on another small mammal the following year, then leaves that host and molts to the adult stage. The following year, it seeks a large mammal, which could be a human, although such ticks seem to be relatively rarely encountered. After feeding, the female lays up to several thousand eggs and then dies. In Yellowstone, adults are active from mid-April to mid-May and are most prevalent in large meadows and in sagebrush-grasslands.*
>
> *The Rocky Mountain wood tick transmits several diseases, including tularemia and Rocky Mountain spotted fever, which was once common in Flathead Valley and the Bitterroot Mountains of Montana but is now much less so because of changes in land use and a shift in the pathogen's virulence. . . . This wood tick can cause a slow-spreading paralysis that is a reaction to the saliva of a feeding tick when it bites on the scalp or back of the neck; the paralysis disappears when the tick is removed.*

He is referring to animal hosts at the end, but tick paralysis can happen to humans, too.

Some kind of insect repellent is a necessity in Yellowstone. The Park Service recommends permethrin against ticks on boots and clothing, and whatever insect repellent works for you as an individual; they normally refer you to the Environmental Protection Agency page "Find the Repellent that is Right for You." In my experience, DEET is best against mosquitoes, although you will regularly hear that it does not work on biting flies at all. I have, in fact, had some luck using it against flies on the northern range, although it has to be kept fresh. Burger, who actually goes looking for these little demons, and counts himself lucky when he wades into a cloud of them, likes citronella—not the candles necessarily, but repellent with citronella oil in it. He has run tests, on the northern range, and citronella performed well. He is also a believer in repellents with aromatic oils in them, oil of peppermint or oil of cloves.

Animals do not have technological fixes, and lack opposable thumbs to apply them anyway, so much of the behavior we see in them every day is in response to the torment of biting insects. Driving through big, open meadows, like Fountain Flats or Hayden Valley, you will see bison rolling on the ground amid a billowing brown cloud. These are dust wallows, and this is one of their favorite activities; the wallows themselves have often been in use for generations, the soil here ground into a fine powder like talcum. The dust makes a coat that biting insects and ticks have a hard time getting through.

The movements of other animals are not random; they are often moving out of still air and into the wind. If they could voice the relief they feel, when the bugs disperse in the breeze, we would hear it. They live most of their existence in complete silence, and it can be frustrating for those of us who wonder about their inner lives. Of this, I am certain: If they celebrated that moment, we would hear it from afar.

Should a tick get on you, getting the thing off requires some care. Among pediatricians and vector-borne disease experts, the traditional methods of removing a tick are no longer advised. They do not want you to burn the tick off, the go-to technique for generations, becoming rare now that no one smokes. They also think you should not try suffocating the tick with petroleum jelly or fingernail polish, which, in addition to being brutal, is too slow and sloppy. Freezing the tick is, too, and has the added drawback that it is crazy to try and nearly impossible. Instead, pull the tick off with a pair of tweezers, grasping as close to the flesh as possible and pulling upward and outward with a steady motion (the Children's Hospital of Philadelphia adds this, for young mothers removing a tick from a child: "Take a deep breath and pretend the tick is just a speck of lint—not an ugly critter with a bloated stomach and writhing legs"). Wash the area afterward, while flushing the tick down the toilet in case it has fight, or bite, left in it. I would keep it in a jar just for fun (I am odd that way); if you think you might want to have it analyzed later, do the same.

There are ways to stay safe, normally, in Yellowstone. Be warned, though. There are bug repellents that will generally do the job—but we have bugs against which no repellent will work.

Chapter Four

One Health, Many Dangers

Zoonotic Agents and Other Small Wonders
The previous chapter looked at vector-borne pathogens. Now, we will take up a new term: zoonosis. Where we were looking at the helpers before, here we will look at those killers that do not need any help as they set out to do you in, although they may be happy sometimes to hitch a ride.

One thing we are doing here is banishing ignorance. In books about Yellowstone animals, it is the big boys and girls that get all the attention, and the little ones get ignored . . . even though you are about as likely to have some kind of dangerous encounter with the little ones as with the large. Therefore, in a book about Yellowstone creatures, the small and less dramatic animals, the ones that get the least attention, deserve an extra helping of our attention, because they are the ones most likely to be at the sending end in the zoonotic exchange. Again (as I will say as many times as seems necessary), it is not just the grizzlies and the wolves that count as important, interesting, and occasionally terrifying Yellowstone life-forms. *Yersinia pestis* is, too, and the other pathogens we are about to meet, as well as the animals they live in. Talk to a bacteriologist sometime, and that person will probably admit (maybe after a few drinks) that *Y. pestis* is just as beautiful as the big animals, too, and certainly as scary.

But back to zoonosis—what is it? It is the formal name for what happens when animals make us sick. It was also the subject of a book by the science writer David Quammen, *Spillover: Animal Infections and the Next Human Pandemic*, a volume published in 2012 that had a second life when the big pandemic happened in 2020. Quammen (who lives, by the way, in Bozeman, Montana, on the northern edge of Greater Yellowstone, and used to work here as a fly-fishing guide), explains the term in a way that highlights the exact points we need:

> *Infectious disease is all around us. Infectious disease is a kind of natural mortar binding one creature to another, one species to another, within the elaborate biophysical edifices we call ecosystems. It's one of the basic processes that ecologists study, including also predation, competition, decomposition, and photosynthesis. Predators are relatively big beasts that eat their prey from outside. Pathogens (disease-causing agents, such as viruses) are relatively small beasts that eat their prey from within. Although infectious disease can seem grisly and dreadful, under ordinary conditions it's every bit as natural as what lions do to wildebeests and zebras, or what owls do to mice.*
>
> *But conditions aren't always ordinary.*
>
> *Just as predators have their accustomed prey, their favored targets, so do pathogens. And just as a lion might occasionally depart from its normal behavior—to kill a cow instead of a wildebeest, a human instead of a zebra—so can a pathogen shift to a new target. Accidents happen. Aberrations occur. Circumstances change and, with them, exigencies and opportunities change too. When a pathogen leaps from some nonhuman animal into a person, and succeeds there in establishing itself as an infectious presence, sometimes causing illness or death, the result is a zoonosis.*

The moment when the zoonosis happens gives Quammen's book its title: It is the moment of spillover. Perhaps the most famous disease that resulted from a zoonosis is AIDS, which spilled over from a chimpanzee into a human in Africa a little over a century ago; perhaps the most infamous is Ebola, which still, and regularly, spills over from an animal that lives somewhere in that same part of the world. The identity of that animal remains a mystery, one reason the disease provokes such fear. There is no telling when or where it might strike next.

When they describe the creatures involved, biologists use special and often surprising language. A metaphor, for instance, inevitably comes up: The germs are said to be "hiding" inside the animal they call home (and "home," for that matter, is a metaphor, too). I am going to let Quammen continue to explain, since he does it so well:

> These pathogens aren't consciously hiding, of course. They reside where they do and transmit as they do because those happenstance options have worked for them in the past, yielding opportunities for survival and reproduction. By the cold Darwinian logic of natural selection, evolution codifies happenstance into strategy.
>
> The least conspicuous strategy of all is to lurk within what's called a reservoir host. A reservoir host (some scientists prefer "natural host") is a living organism that carries the pathogen, harbors it chronically, while suffering little or no illness. When a disease seems to disappear between outbreaks . . . its causative agent has got to be somewhere, yes . . . ? To reside undetected within a reservoir host is probably easiest wherever biological diversity is high and the ecosystem is relatively undisturbed. The converse is also true: Ecological disturbance causes diseases to emerge. Shake a tree, and things fall out.

There has never been a major epidemic that originated in Yellowstone, even though all through the warm months a human river flows over and through the park. One reason, apart from luck, lies in Quammen's last two sentences: No place in the Lower 48 is as biologically diverse and at the same time undisturbed as Yellowstone. Here, things stay put.

There live, in Yellowstone, a variety of life-forms that could be the focal point, any time they "wish," of some mind-blowing zoonoses. They are some of the most dramatic creatures in the park, and yet they never get publicity—or at least not the kind they should. We only pay attention when they cause some human catastrophe. You will not find their pictures on the walls of the restaurants and motel rooms. You will not find them on T-shirts or refrigerator magnets. Their names are spoken in hushed tones. Let us here give them at least a trace of the recognition they deserve without waiting for them to leave the place they belong and turn into a scourge.

Hantavirus is a good place to start. As with so many of these life-forms, Yellowstone surely has plenty of it—hiding here, there, everywhere—and yet we have never had a catastrophic spillover. In the region, it happens all the time.

Orthohantavirus is the formal name for a genus of viruses that the news media and almost everyone else refers to as hantaviruses. There is a tradition in virology by which new viruses are named for the place they are first recognized (as with St. Louis encephalitis), and this genus takes its name from the Hantan River in South Korea, where the first member of the new grouping was identified. It causes a deadly form of kidney failure that added its own contribution to the miseries of the Korean War, afflicting thousands of US and South Korean soldiers—and doubtless, in secret, thousands on the other side. It first came to the attention of most people in the United States when a terrifying outbreak occurred, in 1993, on the Navajo Reservation in the Four Corners region of the Southwest. It was at first a mystery, until a connection was made to the Korean virus, and the scientists investigating the outbreak realized they were dealing with a new variation on an old theme. They tried to follow that tradition and name the virus after the place it was discovered, but the locals complained about every name they came up with, so in frustration, they named it Sin Nombre virus: "no name," in Spanish. The name is unique in virology, an eternal monument to the scientists' annoyance.

The Korean virus attacks the kidneys; Sin Nombre attacks the lungs. It causes hantavirus pulmonary syndrome, a disease that leaves victims feeling as if they are drowning. People who have recovered look back on a nightmare in which every breath required maximum exertion, until, after hours and hours and days and days of it, they were so exhausted they had to fight the urge to give up and sink to the bottom, into death. The cause is what doctors call pulmonary edema: fluid in the lungs, so much that an autopsy of an early victim found that her lungs weighed twice what they should have. Imagine trying to breathe that way.

The first cases in the 1993 Four Corners outbreak happened in May, and by summer, investigators had worked out the cause: It was the urine and droppings of mice, specifically deer mice. When you disturb the droppings, you stir up the virus particles and, potentially, breathe them in. Deer mice are resident almost everywhere in North America, in absurd numbers. Apparently, the virus does not harm them, and we do not get the warning that plague gives us, with stricken rodents leaving their hiding places to die miserably at our feet (Albert Camus's famous novel of the Third Pandemic, *The Plague*, begins this way). Unlike other rodent species, deer mice are not shy, indeed are quite willing to come inside with people and say hello, almost literally, and with their handsome fur and big eyes and ears, they are also cute. No, scratch that—they are adorable. They are everywhere in Yellowstone, including all human habitations. I remember the 1993 Four Corners outbreak vividly because I was working in the park that year, and among my duties, I had to open and clean one of the businesses belonging to my employer. It is an annual ritual in Yellowstone, where there are many hundreds of structures that spend the winter buttoned up tight, completely abandoned . . . to the mice. June 1, 1993, I pulled the heavy wooden shutter off the front door, undid the padlock, opened the door with a loud creak, peered into the dusty darkness, and discovered the leavings of an army of deer mice. I opened a desk drawer, and they had been there. The cash register tray: They had been in every denomination of bills, and the quarters, dimes, nickels, and pennies, too. They had even been in the coin return slot of the vending machine. It looked like they were building a home there. Maybe four weeks later, the news told us exactly how this new Four Corners virus had spread.

We did that every spring, and yet did not get the disease, and of course people who work in Yellowstone still do it, and not just in the park. It helps that there are ways to protect yourself. When entering a newly opened structure, spray liberally with Clorox. Wear a mask, or at least tie the arms of a T-shirt behind your neck with the rest over your nose and mouth, which has the advantage of making you look like an Old West outlaw as long as the shirt does not have unicorns or Taylor Swift on it. A mask is really better.

In the 2017 volume of *Yellowstone Science*, Sarah Haas of the Yellowstone Center for Resources—the office that coordinates scientific research in the park—describes a day looking for hantavirus at the Lamar Buffalo Ranch. If you drive into or out of the park on the Northeast Entrance Road, you will pass the Buffalo Ranch, which was established in 1907 as part of an effort to grow the park's bison herd, and is today used as a field classroom. Haas describes a day spent trapping rodents from the sizable population of such animals living in and among the Buffalo Ranch structures, and searching also for hantavirus. Haas notes the irony. The area is associated closely with bison, and there are regularly hundreds of these gigantic animals visible up and down the Lamar River, but the team of researchers was herding and branding (with ear tags) an animal weighing, in adulthood, maybe half an ounce. She uses the term "megafauna," which biologists use to describe the very largest animals, like elephants and bison. The animals here in their Sherman live rodent traps, baited with peanut butter, were the opposite—but hardly unimportant:

> *Hantavirus is an important disease to understand due to the high prevalence of the virus in rodents such as deer mice (approximately 10-15%, with some areas far exceeding that rate). . . . Understanding the relationship between zoonotic diseases and human-wildlife interactions is a key component of the National Park Service Wildlife Health Program. Although hantavirus is a disease to be taken seriously, the answer is not to eliminate all rodents from the planet, but rather to learn to live with them wisely. This can be especially critical in national parks, where the mission is to preserve ecosystems, including components that could cause human harm. Wise cohabitation with rodents leads to putting up a healthy distance between our living spaces and theirs, including keeping households and buildings safe, clean, and well maintained to exclude rodent entry.*
>
> *There has fortunately never been a documented case of humans contracting hantavirus at Yellowstone. . . . Most visitors who come to Yellowstone may never encounter a rodent such as a deer mouse, or even their diurnal and more respected cousins, the squirrels, and probably leave the park with no sense of missing out. Perhaps the human-wildlife dangers in Yellowstone most people envision (bear*

attacks, bison goring, elk charging, etc.) keep the mind focused on the megafauna that dominate the landscape. When visitors return home, they keep the images and memories of their experiences in play, retelling stories of the large herds and sharp-clawed creatures they observed. Meanwhile, the mice that make up the food chain dance around our houses and whisper stories of their own—how they stole our scraps and made cushions out of our leftovers.

And threatened us with death by drowning, which is basically what happens in a hantavirus case. Part of the irony here, of course, is that when the circumstances are right—or rather, when they are very, very wrong—our megafauna are less dangerous than one of our mice, no matter how much Haas and her colleagues plainly love them.

That has certainly been the case in Yosemite. As with plague, the other "Y" park has had a rougher time than Yellowstone with hantavirus. The outbreak there revolutionized our understanding because it *was* an outbreak; except for the original epidemic in 1993, hantavirus had never appeared in clusters. This time, it certainly did. In the middle part of summer 2012, hantavirus cases started popping up, one after another, and they all shared one surprising feature in common: All the victims had stayed in the iconic Curry Village tent cabins, and all had also, bizarrely, stayed in the new "signature tent cabins" that had opened only three years earlier, called "the 900s" in the language of the hotel employees. It made sense when investigators had all the puzzle pieces. As *Outside* magazine reported, "Unlike the older cabins, which are sided with single-ply vinyl-coated canvas, the signature cabins boasted double-wall plywood construction and propane heaters, making them warmer and quieter than the older units." The double wall had a gap inside, which, combined with the warmth, simply delighted the mice. Hantavirus does not survive long outside the body of its little rodent home; it remains a threat only for about forty-eight hours. "For a human to be infected by such a short-lived virus," *Outside* explained, "you need an ongoing infestation. In Curry Village, deer mice might have occasionally scurried across the floor of the older soft-sided tents. But in the hard-sided 900s they lived in the walls, continually shedding virus." Ten park visitors were infected. Eight of them developed hantavirus pulmonary syndrome, and three of them died. A friend of my family was one of those who died; these things are not always completely abstract.

While there have been no cases of hantavirus infection—that we know of—originating inside Yellowstone National Park, in the surrounding states, hantavirus turns up relentlessly. According to the CDC, between 1993 and 2022, Wyoming had nineteen cases, eight of them fatal; Idaho had twenty-four, five of them fatal; and Montana suffered most, with forty-six cases and eleven deaths. The numbers can go up any time, although for now, in the park itself, the virus continues its uneventful slumber inside our adorable little army of deer mice.

But it is out there, somewhere. Stand on a high spot in the park, or, if you are up for it, a mountain peak. Some are more accessible than others. Bunsen Peak, for instance, is a relatively easy climb; so is Elephant Back—where, however, the bear attack described in chapter 2 happened. Stand there and look out across the great expanse of Greater Yellowstone. And understand that it is out there. All the organisms are. It does not matter how we feel about them, and short of dropping hydrogen bombs, there is little we can do. All the microbes in these pages are as much a part of the park as the bears and the bison. They are stitched into the fabric of the place, and as your vision sweeps the landscape, there they are, somewhere.

There they wait.

Let us now go, gradually, from bad to worse.

There are plenty of life-forms in Yellowstone that will merely ruin your day, without actually killing you, and some of them you should know about simply to avoid the ruining of your day. *Giardia*, for instance, which causes the bottom to fall out of your world. And vice versa.

That actually is an old doctor's joke, and an old ranger's joke, too. I can take neither the credit, nor the blame. *Giardia* is only rarely fatal, but is certainly no joke. So is any travelers' diarrhea, to use the doctor's usual term for it. If you have a health problem in Yellowstone, this will likely be it, or a closely related gastrointestinal problem. It has been the fate of the traveler since there have been travelers.

Why should that be so? It makes perfect sense, if you ponder it. Travel is nothing but disruption, which is the reason we come home more exhausted than when we left. The disruption extends to all things. "Travel disrupts many of the body's natural rhythms, including digestion," according to Kyle Staller, a gastroenterologist at Massachusetts General Hospital in Boston. "Time changes, altered eating schedules, and impaired sleep are all likely culprits, especially in those who already have sensitive guts."

What else? Meals are improvised and hurried. Think of this, for a moment: What was your single most ad-libbed, makeshift meal put together while traveling? Mine: a combined breakfast, lunch, and dinner eaten at Canyon Village in Yellowstone after a day spent on a hike that went wrong and after the restaurants had closed. All I had was five Snickers bars (big ones, too, purchased out of a vending machine—it was expensive). Do that for a week, and you will not even need a parasite to make you feel funny. The parasites are there, though, because the food you are eating is being prepared by someone who often would rather not be doing that, who was just hired for the job and may not be working there a week from now, or a day, or an hour. In national parks and the communities around them, the facilities have a peculiar quality to them that we might call *routine improvisation*. Every year, the business has to begin anew, from scratch, with a new crew hired and trained, and the kitchen brought online from a standing start. Naturally, things go wrong. Don't just take it from me. Here is the CDC on the subject of travelers' diarrhea: It was once "thought to be prevented by following simple dietary recommendations (e.g., 'boil it, cook it, peel it, or forget it'), but studies have found that people who follow these rules can still become ill. Poor hygiene practices in local restaurants and underlying hygiene and sanitation infrastructure deficiencies are likely the largest contributors to the risk."

A range of creatures can be responsible, many of them familiar to you: *Escherichia coli, Campylobacter, Shigella, Salmonella, Cryptosporidium,* and others. Among viruses, Norwalk virus is the major culprit, the main cause of gastroenteritis, or infectious diarrhea. In 2013, Yellowstone and Grand Teton National Parks both had such a bad time with Norwalk that the authorities had to go public with a request that visitors wash their hands more often (seriously, that's what the press release amounted to). That outbreak started by going through a tour group, then clusters of employees, highlighting another reason we get these disorders when traveling: We are jammed together in close proximity with strangers.

It is a Yellowstone tradition, if a minor one. In 1977, a nearly identical situation developed that was centered on Canyon Village. That one warranted a full-bore investigation by doctors who wrote it up in the CDC publication *Morbidity and Mortality Weekly Report*, a kind of "Drudge Report" of contagion and blight. The people involved were caught up in a "We Are the World" togetherness that they would surely rather have skipped. It was the year of the great western drought, the one that baked California so relentlessly that even the largest reservoirs dried out. Yellowstone was short of water, too, so the system at Canyon had to tap a source, Cascade Creek, that was not as pure as the creek it normally used. On the road from Canyon to Norris, you cross Cascade Creek, open to the sky and whatever contaminants might want to get in. People were getting viral gastroenteritis from the water (although intriguingly, some were boiling their water even when it came out of the tap, which would normally be overkill). When investigators looked at how the illness was moving around, they discovered it spreading through dormitories; it was something you caught from the water, but also from your roommate. "Social intermingling among employees in these villages is commonplace," the doctors noted, which reminds me, other life-forms that pose a threat in Yellowstone are the kinds that are transmitted sexually, too. Such organisms must find the summertime park to be a delightful place, what with all those young employees who can scarcely keep their hands off each other.

Perhaps we should not try to cover it all. To continue my campaign for recognition of native pathogens, let us instead return to *Giardia*, which *Clinical Microbiology Reviews* calls "this interesting and enigmatic organism."

Giardia is a genus of parasitic microorganisms that includes the species *Giardia lamblia*, also called *Giardia duodenalis*. It is the one that causes trouble in Greater Yellowstone. For hikers, it can be, and very regularly is, the most important creature in Yellowstone, because it determines what we can drink and how much pleasure we will take in drinking it. On a hot day, when the water supply in the backpack starts to run low, it occupies as big a part of the mind as the bears do, and when the water runs out, the bears get less and less important as thirst grows ever more pressing. *Giardia* then becomes an enemy you can really dislike, more so than any creature in the forest except the biting insects.

It has that devilish quality that microorganisms often seem to possess, as if it were strategizing, demonstrating a satanic cleverness it cannot logically have—or can it? Examining it through a microscope, after it has been stained, the scientist searches for a little smiley face looking back, as if it were, in fact, laughing at us. The stain is highlighting the disc structure *Giardia* uses to attach to the inside of the small intestine. Once in the intestine, the result is giardiasis, the main symptom of which is a nasty diarrhea lasting for weeks, along with abdominal gas, cramps, nausea, and sometimes vomiting. If it goes on long enough, substantial weight loss results, which will sound great to plenty of people, but the people who get giardiasis are typically the kind who do not have much body mass to spare—hikers, that is. When not inside a person or animal, *Giardia lamblia* survives, dormant, inside cysts that can last for months in cold water, and are tough enough to resist chlorine, which will kill the cysts but only when used in adequate quantities. The cysts are, furthermore, everywhere. According to the EPA: "Cysts have been found all months of the year in surface waters from the Arctic to the tropics in even the most pristine of surface waters," and as few as ten cysts will cause an infection.

However, we all carry a galaxy of microorganisms around with us at all times, and *Giardia* is part of the galaxy. The online debate forum *Skeptics* looked at the evidence and decided the threat has been overstated. "The human gut is naturally teeming with microorganisms," they noted. "These are known as your gut flora or gut microbiome. Most of these are bacteria, but quite a few are other organisms, including protozoans such as *Giardia*. Some of these critters in your intestines are beneficial or even necessary for your metabolism, while others may be neutral or harmful. People tend to develop tolerance for their own gut flora, but can get sick from other people's." What may happen is that people "get" *Giardia*, but it does not matter: "*Giardia* is present in about 3-7% of adults in the US, about 30% in the developing world. Among toddlers in the US, roughly a third have it. Most people who have *Giardia* as part of their gut flora have no symptoms at all." There also exists, in the world at large, a universe of false positives and mistaken identities: We get sick from some other cause during a visit to Yellowstone and decide it was *Giardia*, when in fact it could have been anything. *Skeptics* found studies on a literally captive population: "There's also an incubation period, which is the time from infection to symptoms. . . . A 1954 study on prison volunteers showed an average prepatent period of 9 days" (the period between infection by the parasite and the point in time when it shows up in tests), "but there's a wide range of variation, and the incubation period can be as long as months. . . . In summary, if someone gets backpacker's diarrhea while on a weekend backpacking trip, it's very unlikely that it was caused by giardiasis that they acquired during the trip."

The route of transmission is "fecal-oral," as doctors say. You get it when you have sewage in your drinking water, or when you eat without washing after contaminating your hands (although beavers do carry it, and a traditional, jokey name for giardiasis is "beaver fever"). One reason we bury solid human waste in the backcountry, and dog waste, too, is to keep *Giardia* out of the water, among other things. This matter is more important than you may at first think, if you think of it at all (most people don't). The government does not keep records on this kind of thing to the extent you might expect—which has a happy side (can you imagine a representative of the federal government chasing you around the backcountry with a clipboard while you are "performing your necessaries," as Rooster Cogburn puts it?). A 2016 study, however, found nearly forty-five thousand overnight campers in the Yellowstone National Park backcountry during that year, almost all during the warm months, with an average stay of 2.62 nights. The environmental organization Leave No Trace estimates a pound of fecal matter per day, which for Yellowstone would mean one pound multiplied by 45,000 by 2.62, which is an awful lot, concentrated on the trails and around backcountry campsites. The Yellowstone backcountry is *not* heavily used, but that concentration in space and time causes trouble. By definition, all that stuff is being produced in places where it cannot be treated.

Oh, and the urine. Add that, at between one and two quarts per day, roughly. Unless you have a bladder or kidney infection, urine is close to sterile—not quite, but close—so it is mostly harmless, except for one astounding effect you are probably not familiar with. There is a species of squirrel in Yellowstone, the Uinta ground squirrel, that will—once you are finished—very nearly attack the spot on the ground where the urine fell, ripping the vegetation up and eating it right down to the roots, indeed ripping the soil up in their bizarre frenzy to get every drop. Other animals will eat such vegetation, but the squirrels will *freak out*. Watching this spectacle is like watching piranha tear into a cow. Urine has sodium chloride (salt), and that is what they are after, among other things. Still, it is a deeply unsettling sight.

The big problem, though, is the solid waste. The people at Leave No Trace actually ask that you carry it out. If that is too much to ask, which

A gang of Uinta ground squirrels at Mammoth.

it often is, the alternative is to dig a "cathole" six to eight inches deep and two hundred feet from the nearest water source, trail, or campsite. Carry a plastic trowel to dig with; the ground in Yellowstone is mostly rocky glacial miscellanea, a horror to dig in, and impossible with fingernails alone.

In the backcountry, disinfect water before consuming it. I agree with the people at *Skeptics* that the danger is overstated, but it is better not to take the risk, and easy enough. The simplest solution is water purification tablets, like those made by Micropur, which release chlorine dioxide. Even old-fashioned iodine can work; the trick is to carry a package of Kool-Aid to kill the taste. As a related aside: A common habit of hikers everywhere is building an emergency kit that would do the Navy Seals proud, and then never looking at it again, or at least not until it is too late to address any shortcomings. For instance, I felt like a genius when I managed to acquire a doctor's penlight, the kind they use to assess pupil dilation; it was the perfect emergency flashlight. I felt less like a genius when I at last needed it, years later, and discovered that the batteries had grown blue-green fur. Micropur tablets will not grow fur. Just remember to check the expiration date. Or you can use a filter, or boil the water. Even plain bleach will work, as long as you use enough. People tend to overthink these things.

Finally, at the start of this section, I said that we would go from bad to worse where these creatures are concerned. What else counts as "bad"?

Yellowstone is not as accomplished as the parks in the desert Southwest, where these next organisms are concerned, but we do have insects and arachnids that will take a piece out of you, and not just as disease vectors—they can do some harm without the germs to help them in the effort. We have leeches. They do not transmit disease the way mosquitoes do, and so are mainly regarded as a nuisance—unless you are a fish, in which case they are dinner. Having found a human, they attach themselves to the flesh and inject an anticoagulant, then drink their fill; getting into the water up to and past the waist gives them an invitation. The real experts on leeches may be fly anglers, who have "flies" that imitate all the different varieties ("Whether you are fishing in winter or summer," says one Jackson, Wyoming, fly-fishing shop on its website, "in high or low water, it's always leech season").

An unusually well-fed black widow found in a VRBO down the valley from Gardiner.

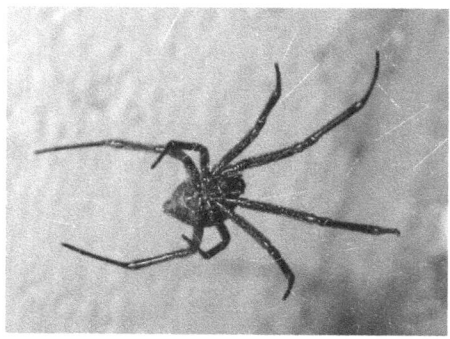

A more svelte black widow.

The two dangerously venomous spiders that are a worry in North America, black widows and brown recluses, do not seem to occur inside the Greater Yellowstone national parks (and no scorpions, either). Brown recluses, a southeastern species, would appear to be a geographic impossibility, but people living in the lower elevations outside the parks do report black widows in places like garages and basements, and I have seen them myself in the Yellowstone River valley, downstream from the North Entrance. The species is *Latrodectus hesperus*, the western black widow spider, and the identification technique you likely have heard of—glossy black body with the famous red hourglass on the underside—is correct and works well. The species seems to be fairly common in the rest of Montana, so if your VRBO has a basement, they may well be down there, waiting for you. We also have wolf spiders, which live in burrows and stalk insects as a good predator should, mostly at night. This habit limits the number of run-ins it has with humans, but it does produce venom, and the bite can be painful to those sensitive to it.

Finally, we have bumblebees of great charm. Their sting is not to be mocked. A bumblebee's stinger is not barbed the way a honeybee's stinger is, so a bumblebee can sting again and again without injuring itself, like an enthusiastic soldier in bayonet drill. A friend was stung one day at work at Mammoth Hot Springs in mid-July of 1994. It was painful merely to watch. One side of his face was perfectly normal, while the other looked like a red throw pillow. We could have gathered his hair to make tassels.

What, then, counts as worse? And worst?

We surely have anthrax and tetanus, although noting that we do is the equivalent of saying, "Don't tell anyone, but we have water and soil." The creature that causes tetanus is close to universal. You do get more of it in a wetter, warmer place than Yellowstone, but nevertheless it is here, and it is so nearly universal that you may be carrying some around in your intestines—harmlessly. Tetanus is caused by a bacterium, *Clostridium tetani*, and actually has nothing to do with rust. Many (perhaps most) of us retain a belief from childhood that the bacterium grows in the rust on old fishhooks and nails. But it is the puncture wound that fishhooks and nails cause that is to blame, not the rust itself; sharp pieces of steel like that make wounds that close up and create the oxygen-free environment the bacterium needs to thrive. Inside such a wound, it makes a poison, tetanus neurotoxin, that is the cause of the harm. This particular poison trespasses on the realm of science fiction (and so it makes sense that we have it in Yellowstone). It is one of the most ferociously poisonous substances known to humanity; only botulinum toxin is worse. Fatal doses are measured in nanograms—that is, a billionth of a gram. It kills by more or less literally torturing the victim to death, keeping in mind that the word "torture" comes from the Latin word for "twist." The toxin causes muscles to spasm all over the body (hence, the lockjaw by which tetanus is known), and the spasms can be powerful enough to break bones. It may be more of a threat to people who work outdoors: In 1958, tetanus killed a National Park Service employee at Sequoia and Kings Canyon, a man named Charles Wallace who was part of the crucial maintenance division that keeps the parks running. Wallace died when tetanus developed after a yellow jacket sting, so any puncture wound can be a danger.

Anthrax has more the flavor of Yellowstone about it, given what it is and what it afflicts. It, too, is caused by a bacterium that is universal in soil around the world, *Bacillus anthracis*, and while it can be found on every continent except Antarctica, it is more common in some places than others, to the extent that some areas are considered "endemic" for anthrax. It most often affects big grazing animals of the kind we have so many of in Greater Yellowstone. It is rare in the United States, and looked at on a county-by-county basis, it is almost entirely absent from the Northern Rocky Mountains—almost. One county has reported cases, quite a few, since 2000: Gallatin County, one of the five that surround Yellowstone National Park. Bozeman is the county seat. On a map of anthrax endemicity, most of the West is blank, except that one county, blazing red in a sea of white.

You may have even heard of the place where it all happened: the Flying D Ranch, owned by media mogul Ted Turner. If you leave the park through the West Entrance and head north from West Yellowstone, you will drive past it. The Flying D is operated partly as a working bison ranch, partly as a kind of 113,000-acre private nature preserve; it grows bison as livestock, a significant business in this part of the world. In 2008, a small catastrophe happened: Not quite three hundred bison from the ranch herd of just over four thousand died of anthrax. For three weeks in the summer of that year, ranch managers, ranch hands, state officials, veterinarians, and others were in the middle of a bad movie. They buried some of the animals, burned others in an incinerator the size of a boxcar, and disinfected "hot spots" in the soil—places where an animal had laid down and died—by liberally dosing them in bleach. The Flying D was quarantined, roads in and out closed. The state epidemiologist's office announced that meat would be kept out of stores (although such meat has to be well undercooked to pose a hazard). Stricken animals were treated with antibiotics, and the last death was August 24.

It happened again in 2010, when a single bison calf broke with anthrax. At this later date, the Flying D herd had been vaccinated against *Bacillus anthracis*, and ranch hands carried portable kits to test for anthrax. Vaccination takes place in the autumn, though, and the sick calf had not been vaccinated yet. Furthermore, the vaccine apparently does not always work, or wears off.

There were no more big outbreaks among the domestic bison herd. However, during the 2008 calamity, anthrax was suspected in the death of a black bear that turned up deceased on the ranch. Then, later that year, the outbreak claimed new victims, as the Bozeman *Daily Chronicle* reported: "Naturally occurring anthrax in the Spanish Creek, Elk Creek and Cherry Creek drainages near Ted Turner's Flying D Ranch has killed two deer and is likely responsible for the death of 14 elk, a Fish Wildlife and Parks official said Tuesday. The animals probably became ill after coming into contact with anthrax spores in the soil." Investigators later ran the total of anthrax-killed elk up to forty-three. The same investigators were able to zero in on the parts of the ranch that were most dangerous (the "foci," same as with plague); the ranch hands kept the animals away from them. They also identified the climatic events that triggered the eruption: "The outbreak began after a midsummer rainy period in a hot, dry summer, classic conditions for anthrax epizootics"—an epidemic, but for animals. "Bison deaths were concentrated in two large pastures in the southwest of the ranch, and a third northern pasture erupted late in the outbreak." In 2010 these same investigators found a pair of bison on the ranch that had been sick with anthrax, and then were killed and eaten by wolves; one of the bison had not been vaccinated yet, but with the other, the vaccine had simply failed. The wolves, they reported in the *Journal of Wildlife Diseases*, were unharmed, and the researchers were left to ponder whether this might be a strategy wolves exploited in the long stretches of prehistory, and continue to exploit: Look for the sick bison and pick it off instead of the healthy ones, which are prohibitively large and dangerous. (And by the way, the researchers gave their report a delightfully tasteless title: "Dances with Anthrax.")

The disease, as with so many illnesses caused by bacteria, is no longer the killer it once was. Today, it is mainly a threat to animals, most of which are not vaccinated. *Bacillus anthracis* makes spores. Animals get the disease by swallowing or breathing in the spores while grazing, or from other animals that already have it, which is how people usually get it. It causes ugly black skin lesions that give the disease its name, which comes from the Greek word for "coal" (like "anthracite"). It also has intestinal and inhalation varieties, the latter especially dangerous. Both do what diseases always seem to do: They look just like the flu, until you keel over. The spores can survive for decades and are tough to destroy; because they are also easy to aerosolize, they have been of great interest to nations looking to develop bacteriological weapons. The 2001 anthrax letters incident that happened in that nervous period just after the September 11 terrorist attacks involved aerosolized spores; of twenty-two people who developed anthrax, five died, in part because they had inhalation anthrax.

Animals that graze are at greatest risk. We have a great many in Greater Yellowstone, and not just the big, famous ones like bison and elk; a range of other, smaller animals graze, too. Animals develop the disease three to seven days after exposure, and can die in hours. Perversely, one clue that makes a veterinarian suspect anthrax is the carcass of an animal that appears to have died of nothing. You would probably not mess with one anyway, but the way to avoid anthrax is to stay away from dead animals.

It is, however, another disease that you would not change your travel plans over. In July 2011, a sixty-one-year-old man was hospitalized for pneumonia in Minnesota but, as it turned out, was actually sick with anthrax. He and his wife had just completed the kind of road trip of the western United States that so many people make as part of their Yellowstone vacation. They had a wonderful trip, as the CDC reported after its investigation: "They drove through North Dakota, Montana, Wyoming, and South Dakota, where animal anthrax is sporadic or enzootic"—meaning endemic in the animals. "While traveling, they walked in national parks, collected loose rocks, and purchased elk antlers. On July 29, they drove through herds of bison and burros, frequently stopping while animals surrounded their vehicle." They spent time in both Yellowstone and Grand Teton National Parks. Then, traveling east, the husband landed in the hospital with what doctors thought was pneumonia. The CDC and even the FBI investigated his every move and found nothing.

It can happen any time—let's say to one of our big grazing animals, like a member of the free-ranging Yellowstone bison herd. Out of nowhere, one of them will keel over like a torpedoed ship, then another, then another, of anthrax. If no one notices, however, no one with the knowledge to ask the right questions and get researchers into the field to run tests, nothing will happen. The tree will fall in the forest, and it will not make a sound. And, given the way people freak out over these matters, perhaps that's for the best.

Which leads me naturally to our next Yellowstone pathogens, over which people *really* freak out. I can hardly blame them, when a search for the next one, *Naegleria fowleri*, sends you to the "brain-eating amoeba" pages.

That is what they call it, even level-headed scientists who are, I suppose, as fond of drama as anyone, when they are the stars. *Naegleria fowleri* is another microorganism, this time an amoeboflagellate—that is, an amoeba with a whip-like flagellum that it uses to move around. Most of the time, it earns an honest living as a hunter, attacking and consuming bacteria. It lives in warm water. In cooler conditions, under 50°F, it forms cysts that can withstand the cold until the temperature improves. When it does, *N. fowleri* leaves the cyst and enters an active, feeding stage. It seems to thrive best at about 108°F, a few degrees warmer than people normally prefer to set their Jacuzzi, and it continues to do well up to 115°F. It cannot stand saltwater, and is sensitive to acidic conditions and drying. It wants warm freshwater, including lakes and rivers as long as the temperature stays up. It also does well whenever local conditions spike the thermometer, such as in warmwater discharge from power plants, in residential or commercial water heaters, in piping connected to such heaters, in geothermal well water, or in the kind of swimming pool that annoys the county health authorities, with the water all green and full of mosquito larvae. The ancient Roman baths in Great Britain have been closed to swimming since the 1970s because of *N. fowleri*. It wants warmth, but it also hungers for the bacteria it consumes for a living. As you might guess, the hot springs and other thermal water so plentifully available throughout Greater Yellowstone is an open invitation to *Naegleria fowleri*.

In that active, feeding stage, it does a remarkable and rather gruesome thing: It swims up the human nose and follows the olfactory nerve into the brain. It just bores right in. The result is a disease called amoebic meningitis, more properly primary amoebic meningoencephalitis. It is also called naegleriasis, infection by *Naegleria*, and it is now that the "brain-eating" part happens: The creature ceases to feed on bacteria and switches to brain tissue. The symptoms are the kind you would expect from an attack on the brain; they are meningitis-like, among them headache, nausea, fever, confusion, vomiting, hallucinations, and seizures. The disease progresses into a kind of crescendo in less than a week, and is almost always fatal.

Naegleria fowleri occurs all over the world but is thought to have evolved in North America, where most of the reported cases in the brief history of the disease have occurred. Rare though it is, it is a perpetual worry to public health authorities in places like Texas and Florida, where, when summer heat waves get serious, people all over these states get into whatever water is available as a way of escaping the heat, no matter how skanky that water might be. The brain-eating amoeba is not on people's minds at such moments.

It appears to be another native Yellowstone life-form. Researchers in the early 2000s used genetic technology to search for it in Yellowstone and Grand Teton waters. The investigators took samples from a wide range of thermal water and found *Naegleria fowleri* in the Boiling River, Seismic Geyser, and what they referred to as the "Mallard Lake Trail site." The Boiling River is on the Gardner River below Mammoth Hot Springs; Seismic Geyser is downstream from Old Faithful, between Morning Glory Pool and Biscuit Basin in the Upper Geyser Basin; and Mallard Lake is a hiking destination three miles from Old Faithful, and the trail there passes thermal activity where the researchers took their samples. They were only really confident about the Boiling River, but they found traces of *Naegleria*-like organisms all over the place.

The one location everyone is certain about, and has been for some time, is actually between Yellowstone and Grand Teton National Parks, in a stretch of US 89 that is administered as the John D. Rockefeller, Jr. Memorial Parkway. It's a peculiar thing, that parkway. It appears, at first, to be the result of a cockeyed effort to flatter a billionaire. More charitably, it was an effort to connect the two parks, Grand Teton and Yellowstone (it includes twenty-four thousand acres of surrounding terrain). The area hosts more than a million visitors a year, maybe closer to two million, and most have no idea they are visiting, or even that the place they are visiting exists. One feature that is part of the parkway is a thermal area to the west of the road. It includes Huckleberry Hot Springs and Polecat Creek Hot Springs, a second thermal zone. They are in the middle of nowhere, at the end of a hike, but are well known locally because of an old tradition: using hot springs as natural Jacuzzis. It is called hot-potting, and you may even have been planning on trying it out, so it deserves a look.

Hot-potting has, for many decades, had a single legal outlet in Greater Yellowstone, one that was known literally throughout the world: the Boiling River. As noted above, it is in the far northern section of the park, below Mammoth Hot Springs on the old road to the North Entrance. It is a peculiar phenomenon, a creek that emerges from a cave to run perhaps two hundred feet, then empties into the Gardner River. The water is flowing down from the hot springs above, at Mammoth, and so is warm—too hot to get in, actually. Once it entered the Gardner, however, it mixed with the extraordinarily cold water of a mountain river to make a natural hot tub that was just right.

Notice I switched my verbs to past tense. In June 2022, a flood of historic violence turned the Gardner into a raging nightmare. It was like watching a relative who should stay away from alcohol getting drunk at Thanksgiving. In places, it eroded the bank, and in other places, it deposited debris, and it performed both actions with extreme vigor. At the Boiling River, it filled in the junction of the two watercourses with a mix of gravel and rocks and even a few boulders. Downstream, it chewed sections out of the access trail, and it destroyed the road; the North Entrance Road now follows a completely different route. So, for the time being, the Boiling River is not what it was. Getting in is technically illegal—but in fact, there is no "in" to get into. You would need to find a way to bathe in rocks and gravel. The Gardner will probably reroute itself eventually and carry away enough of the offending debris to reopen the area to swimming and soaking, but for now, it effectively does not exist.

There are commercial operations that offer thermal water in a slightly more civilized setting (Chico Hot Springs Resort is a famous one, located in Paradise Valley, along with the newer Yellowstone Hot Springs near Gardiner). People do like the wild and free version, though. There are springs inside Yellowstone National Park that fit that requirement but are illegal to enter; they are also local secrets, the kind of thing that comes up in conversation, eventually, when you hang out here for decades. That leaves Huckleberry Hot Springs and Polecat Creek Hot Springs—which are also closed now. Back in 2014, the superintendent of Grand Teton National Park decided to start enforcing an old NPS regulation that prohibits "swimming or bathing in a thermal pool or stream that has waters originating entirely from a thermal spring or pool." The superintendent, David Vela, said he was making this change to let the springs recover from the trampling that has gone on for decades. "We are responsible," he said, in a press release, "for applying the best stewardship practices that allow for eventual restoration and conservation of these unique resources under our care." At the same time, hot-potting "will still be permitted in any creeks or pools not solely of thermal origin. Features that remain open for public use are only those warmed by the runoff from nearby hot springs, such as Polecat Creek itself, which provides a desirable environment in which to soak." The superintendent was trying to keep the public happy and do his job at the same time—not always an easy task in this business.

There is one other problem with these locations—yes, the *Naegleria*. As noted, that earlier study found it in the Boiling River, and the flood will not have washed it away. In the 1980s, scientists collected *Naegleria fowleri* in Huckleberry and Polecat Creek Hot Springs; when I worked in the park in the 1990s, in conversation, "Huckleberry" and "meningitis" went always—if vaguely—into the same sentence. They found it again in 2002, and more recently, too. These studies are complicated. It is certainly a cliché to talk about needles and haystacks, but finding an amoeba-size microorganism in a pool the size of a haystack is worse. Researchers therefore look for whatever they can find, including traces of genetic material. The technology continually improves, and because of that improvement (since, for instance, the study from the early 2000s that found *Naegleria* in the Boiling River), we now finally have some definitive answers.

In 2024, a study was published in the American Chemical Society's journal *ACS ES&T Water*, produced by researchers mostly from the US Geological Survey and the CDC, that appears to have resolved the mystery. The researchers looked at the Huckleberry and Polecat springs in the Rockefeller Parkway, and also Kelly Warm Springs, a site in the far southern part of Grand Teton National Park. They took samples from high-volume and low-volume water; they also sampled biofilm (the slimy film bacteria make that adheres to surfaces) and the sediment on the bottom. They ran qPCR, real-time polymerase chain reaction, to amplify genetic material in the water enough to analyze it, and also cultured samples to see what they might find that way, before running qPCR on them. They sequenced the DNA. They took samples over the course of a year, during different seasons. They did some other stuff, too—basically everything but turn the springs upside down and shake them. The result? "With larger sample sizes, multiple sample types, and updated analytical methods," they wrote, "the present study provides the first detection of the presence and viability of *N. fowleri* in surface water, sediment, and biofilm samples from GRTE [Grand Teton] hot springs," which here includes the Rockefeller Parkway. "These results provide new insights into the distribution of both pathogenic *N. fowleri* and nonpathogenic *Naegleria* in natural thermal water systems in northern latitudes." They were able to distinguish between dangerous *Naegleria* and harmless *Naegleria*, and were able to find it in places where previous studies had shown nothing. Their findings are conclusive.

I have never seen the appeal of hot-potting. Thermal water is a living, breathing rainbow of every imaginable life-form that will thrive in warm water, like *E. coli*, or *Clostridium difficile*, or *Salmonella*, *Campylobacter*, *Shigella*, hookworm, roundworm, *Entamoeba histolytica*, *Giardia lamblia*, *Cryptosporidium parvum*, and a wide variety of others—the Latin and Greek dictionaries can hardly contain them all. Yes, they *are* all in there, even though you can't see them. I never could figure out why people want to sit in that. The real danger, though, may lie simply in the walk to the hot pot. In 2016, a young man fell into a spring at Norris while looking for a natural Jacuzzi. By just the next day, there was nothing left. His body, long since dead, had literally dissolved.

But at the same time, not *every* single spring is off-limits, and to each his own. Writing this, I have been continually reminded of a passage in Edward Abbey's 1968 book *Desert Solitaire: A Season in the Wilderness*.

The book takes the form of a diary tracking a summer Abbey spent at what was then Arches National Monument in Utah in the late 1950s. Paperback copies are a common sight around dormitories and cabins where employees live in Yellowstone, as they were when I worked here; in fact, my first summer, a copy was in my luggage. Reading it was the major reason I came to Yellowstone in the first place. It was the inspiration. *Desert Solitaire* is a complex book, part serious philosophical inquiry, part political diatribe, but one theme Abbey returns to repeatedly is the sad way we limit ourselves in the wilderness, and the national parks especially. We allow ourselves to be satisfied with civilized comforts, when a greater, bolder experience is there for the having. He writes this: "A venturesome minority will always be eager to set off on their own, and no obstacles should be placed in their path; let them take risks, for godsake, let them get lost, sunburnt, stranded, drowned, eaten by bears, buried alive under avalanches—that is the right and privilege of any free American."

I do not disagree—while recognizing that administrators have to do what they have to do. National parks have safety rules for many reasons, a good one being the simple reality that most people are not equipped to take any risks at all; travel in the outdoors requires a wide variety of skills, some of which take years to acquire. Still, all this time, I have felt like I am betraying Abbey's spirit. We will return to this topic in the last chapter. For now, know this: No one has ever had naegleriasis in Greater Yellowstone, or indeed in the whole Northern Rockies. The CDC tracks cases, of course, and between 1962 and 2023, there were only 164 diagnoses of primary amoebic meningoencephalitis in the United States. Draw a line from northern California across the continent to northern Virginia, and almost all of those cases happened below that line, where the water is warmer. In Wyoming, Montana, and Idaho, there have been zero cases. Indeed, zero cases in those states and in the Dakotas, Nebraska, Colorado, Utah, Oregon, and Washington, the whole northwest quadrant of the nation. Yours won't be the first. Too much has to go wrong for this normally inoffensive creature to go on its rampage.

Still, and again, it is out there. And so—to somehow go from worst to even worse—is rabies.

Yes, it still exists, although it is rare enough in North America that nearly no one knows what it actually is. Like naegleriasis, it is a kind of encephalitis, an inflammation of the brain, caused in this case by a virus, *Rabies lyssavirus*, that the victim acquires most often from the bite of a sick animal. It begins with a fever and headache, and with tingling, prickling, and itching around the site of the original wound; after that, it gets more alarmingly strange in ways that suggest, correctly, that the central nervous system is being victimized—or more accurately, brutalized. Sufferers become irritable and aggressive, emotions that may take a physical form, the victims thrashing around in extreme agitation. Their thinking becomes confused and bizarre, with hallucinations. They experience muscle spasms and become contorted into tortured postures. They suffer seizures, with visible convulsions. They grow weak, and parts of the body are paralyzed. Lights, sounds, or other sensations become a torment. Strangely and distinctly, they develop a fear of water, literally: Show the victim a glass of water, and the person recoils in actual terror. It sounds impossible, but that is what happens; it gave the disease its old name, hydrophobia. Just as strangely and distinctly, victims produce so much saliva that, unable to swallow it, they literally foam at the mouth. In poorer countries, these symptoms enable doctors to spot the disease reliably when a laboratory test is harder to come by.

Here is another creature (to the extent that a virus is a creature at all) that seems to take its inspiration from science fiction. In *Shadows of Forgotten Ancestors*, a book he wrote with his wife, Ann Druyan, the physicist and old-time public television celebrity Carl Sagan captures this quality well. He writes about how even the simplest life-form will develop devilishly sophisticated ways to reproduce itself:

> *Of course, predators need not be bigger than their prey. Disease microbes can be formidable predators—not only attacking and eventually killing the organisms that bear them, but also taking over their hosts, changing their behavior to spread the disease microorganisms to other hosts. One of the most striking examples is the rabies virus. On being injected into the bloodstream of a placid, people-loving dog, they head straight for the limbic system of the dog's brain, where the*

control buttons for rage reside. There, they set about converting the poor animal into a marauding, snarling, vicious predator that now bites the hand that feeds it. Rabid animals are afraid of no one. At the same time, other rabies viruses are dispatched to inactivate the nerves for swallowing, to put the saliva-manufacturing machinery into overdrive, and to invade the saliva in huge numbers. The dog is furious, although it has no idea why. A pawn of the viruses within it, it's helpless to resist the impulse to attack. If the attack is successful, the viruses in the dog's saliva enter the bloodstream of the victim through the lesion or laceration, and then set about taking over this new host. The process continues.

The rabies virus is a brilliant scenarist. It knows its victims, and how to pull their strings. It circumvents their defenses—infiltrating, outflanking, accomplishing a coup d'etat within beings so much larger, you might have thought them invulnerable.

Sagan is aware that the virus is unaware; it is the dumbest creature imaginable, a mere packet of genetic material needing a cell to hijack, using the victim's cellular machinery to replicate itself. Nevertheless, it is amazing. The thing appears to have a satanic intelligence.

Everything about it is creepy. Not every pathogen uses the bloodstream as its main highway to reach the places it wants to exploit. Plague bacteria, for instance, travel along the lymphatic system, with the buboes being ruined lymph nodes. Rabies virions—virus particles—target the nervous system. It starts at the point where it entered the victim's body, and from there, it moves from neuron to neuron, in a way that is deceptive—and again, with the science fiction. As *Scientific American* explains, "Whereas many virus species replicate so rapidly that they force the infected cell to burst open, releasing the virions into the space between the cells, the rabies virus strictly regulates its reproduction—producing just enough daughters to keep moving on. That way, it refrains from causing so much damage that it alerts the immune system. Instead, it leaves the host cell intact and crosses a synapse to a new upstream neuron. That sneakiness is one reason the disease has such a long, symptomless incubation period, typically one to three months in humans." The ultimate destination, as Carl Sagan noted, is the brain and the salivary glands. The victim is driven mad, and driven to bite and distribute what we might think of as weaponized saliva. It can take as little as four days, but there are cases on record in which the incubation period—the virus's trip from the bite to the brain—has lasted six years.

Again, we are talking about zoonotic life-forms in this chapter. Rabies does involve a vector; it is an instance in which the vector is not a biting insect, but a warm-blooded animal, normally a dog, but possibly a skunk, raccoon, fox, or another creature with sharp teeth. Other animals can get rabies, but pose less of a threat; rabies specifically makes animals furious, and a furious grazing animal is more likely to attack in other ways, by kicking or goring. You need the bite to get the disease, and so herbivores are also less likely to give it to each other in the first place. Rabies is now rare in the United States, quite rare, because we vaccinate our pets, and especially our dogs. In the developing world, it remains a horrible problem, with almost sixty thousand cases thought to occur every year—and note the word "thought." The number is likely significantly higher from unreported and unidentified cases. If you want to see what rabies looks like, and feel like you have the stomach for it, look at the videos on YouTube of rabies victims, including a series posted by the British medical journal *The Lancet*. The victims exhibit the hydrophobia reaction (one reason *The Lancet* put them there), and literally foam at the mouth.

In this country, we once lived with it, too. Note how often it pops up in our popular culture, especially older popular culture. Think about novels—and you may have already thought of one of these examples. In the 1956 novel *Old Yeller* by Fred Gipson, later the hit Disney film, the dog Old Yeller saves the family from a rabid wolf and later, tragically, develops rabies himself, so that the boy at the center of the story has to shoot him (and hence, the words to the country and western song: "She never cried when old Yeller died / She wasn't washed in the blood of the lamb"). In Zora Neale Hurston's *Their Eyes Were Watching God*, the central character, Janie, is saved from a rabid dog by her husband, and when he develops rabies, she has to shoot him. In Harper Lee's *To Kill a Mockingbird*, the little town goes on high alert when a dog turns rabid, and the sheriff asks the hero, Atticus Finch, to shoot it, because of his expertise with a rifle. He does so, in a memorably gripping scene. All these scenes are gripping, and notice that, judging by their fiction, our forebears went straight to their firearms when rabies got loose.

A more recent appearance of rabies in popular culture is worth including: An episode of *House M.D.* that aired in 2005 centers on rabies as the final diagnosis. In a way, this makes perfect sense, as the series is about an obnoxious, highly eccentric, impossibly brilliant doctor who diagnoses impossibly rare diseases. This was almost a running gag in the series, so that the fan website, the House Wiki, has a Zebra Factor that depends on how improbable the diagnosis in each episode is (this one actually only had a Zebra Factor of 7/10).

Rabies, in the United States, is about as common as *Naegleria* infections, almost exactly. Again, between 1962 and 2023, there were 164 cases of naegleriasis; during a comparable period, 1960 to 2018, there were 125 reported cases of human rabies in the United States. Only 89 were actually acquired here; there is so much international travel now that people commonly bring things like rabies home after having been bitten by a dog somewhere in the developing world. Even pushing the earlier date back to 1938, into the Great Depression, the number of cases only goes up to 588. It is true that it remains almost universally fatal, once the symptoms have advanced to a certain degree. After exposure—say, after a bite from an animal that could be rabid—treatment with post-exposure prophylaxis (PEP) is part of what has brought the death toll down so far in the United States. This consists of a thorough washing of the wound, a dose of human rabies immunoglobulin, and four doses of the vaccine. That post-exposure course you may have heard of, with the seventeen intensely painful injections into the stomach muscles, was something they did before 1980, although so few people know about this change that they assume a scratch from a strange dog is the knell of doom.

Greater Yellowstone has most of the animals that typically function as the vector, although any mammal can get it. As near as I can determine, there has never been any kind of program to test large numbers of animals in Yellowstone National Park for rabies. When incidents occur, with park employees and visitors being scratched or bitten by wild animals, there will be a test if the animal can be gotten hold of, dead or alive. A coyote that bit a skier at Canyon in 2020 tested negative, an incident we will return to later. In Glacier National Park in 2019, a bat that scratched a resident of the St. Mary area tested, in fact, positive, and the victim underwent the post-exposure prophylaxis treatment, which has become a more or less routine matter in such cases. It is not otherwise the job of the National Park Service to chase wild animals around with medical equipment, waving needles and syringes.

It very much is the job of the health authorities in the surrounding states—once a suspect animal is in hand—and they find rabies with some regularity. Just to give some idea of the numbers, tests are run on suspects by the Montana Department of Livestock (a big deal, in Montana—kind of like the Department of Defense but for Montana). Between 2015 and 2019, they found rabies in three skunks, one cat, and sixty-seven bats. This count included bats on the north boundary of Yellowstone National Park, in Gallatin and Park Counties. In Wyoming, testing is done by the Wyoming State Veterinary Laboratory at the University of Wyoming in Laramie. In 2023, they found six positive skunks and fourteen positive bats, including three bats in Teton County. Idaho is comparable, with sixteen positive bats in 2023 and no other animals.

Yellowstone National Park does have raccoons and skunks, although both are so rare inside the park boundaries that I have never seen one. However, Gardiner, Montana, is right on the park line—parts of the town actually jut over it, depending on how you define "town"—and Gardiner began seeing more skunks starting about fifteen years ago. At our home there, we had one under the tool shed starting in late spring 2024 and ending around the time I am writing this, as best as we can determine. She took up residence before we could very easily do anything about it, and then she became like the eight-hundred-pound gorilla. We don't want to get too close, although they generally only spray when you really threaten them by, say, accidentally stepping on one, which I did once while fishing in California, and she did not like it, as I discovered (and by the way, the thing about tomato juice taking off skunk spray is a myth; hydrogen peroxide mixed with baking soda and soap is the solution). Our Gardiner skunk is a female, but has not yet had kits. That is the word for them, and they make a sound like a baby guinea pig. All through the summer of 2024, I quizzed people and watched our own animals. There had been at least three skunks living inside Gardiner, but one was killed early in the summer by a local after spraying the local's dog twice. That left a female and a male, and I expect kits soon. Don't tell anyone in town, but I find them cute, kits and adults both. Rural Montanans would never tolerate such an attitude.

We have foxes in Yellowstone National Park, red foxes specifically. We have plenty of other animals that can get rabies. One of them we have in great numbers; it is a Yellowstone animal that visitors often fear as much as they do the bears, and really should not. I wish I could talk them out of it, because this is their home as much as it is ours, and probably more.

Yellowstone is bat country.

There is a tendency to be alarmed by their presence, but you need not be. The bats in Yellowstone would only have an opportunity to bite you under bizarre circumstances, and they do not want to, anyway. Vampire bats do not live anywhere near Yellowstone; they are a warm-weather animal, and the closest they ever get to here is the extreme southern part of Texas. If one of our bats did bite or scratch you, you would have to undergo the post-exposure routine, but it is not likely to happen. They do live in human structures in the park, especially in the older hotels; they were in the shack I worked at in the village at Old Faithful, a quaint old structure that dated to the 1920s.

Our experience, actually, is instructive. We have the saying "bats in the belfry" for a person whose sanity we want to question, because it is the head of the person that is the problem. The belfry is the architectural equivalent, the place where the bells are at the top of the structure, open to the air. In the hotels and the other large, older buildings, bats are in the attics, as they were in our little structure. We happily went our separate ways—except one time. A bat got loose in the main room, the one we called the office, and circled around a friend of mine, around and around, like a badly damaged plane orbiting an aircraft carrier. My friend had the composure to open a door, and the bat—no doubt happily—escaped. Or the incident recorded on the Yellowstone Park employee Facebook page, in which a Park Service ranger, on duty at one of the entrance kiosks, felt and heard a *thwap* on his head. He took off his "flat hat," his service Stetson, and discovered a bat clinging to the brim. I get the impression that in such situations, the bat will seek any port in a storm. The sun was out, and they are nocturnal; when a bat finds itself outside in daylight, perhaps against its will, it will seek any refuge that makes even marginal sense. The ranger coaxed the bat off, which can be difficult. This is the moment when a bite is a possibility. If it happens inside the park, ask for help from an employee. That person is likely to either know what to do, or know how to contact someone who does.

People think of them as "flying rats," which is not right in a bunch of ways. The average Yellowstone bat is not that big. They appear much larger than they are because of their wings; they are unintentionally bluffing us. A common bat in Wyoming, maybe the most common, is the little brown bat, which sounds like the title of a children's book, but that is what they are called: the little brown bat, *Myotis lucifugus*. It is, well, little, with a body length typically a bit over three inches and a total

The little brown bat. NATIONAL PARK SERVICE

weight in the range of a quarter or a third of an ounce. Rats of the larger species (*Rattus norvegicus*, the brown rat) may be ten or eleven inches long and weigh over a pound. More importantly, we have so many rats because of the way they reproduce. An important concept in biology and ecology is the r/K-selection theory. Briefly, how does a species reproduce? If it creates a large number of offspring and puts little effort into raising them, that is the r-selection strategy; if it does the opposite, producing few and putting a great deal of energy into every one, and also generally living much longer, that is the K-selection strategy. In one, offspring are cheap, in the other expensive, which will sound familiar if you have children: Humans, like grizzly bears and bald eagles, are K-selected. Here is a crucial difference between rats and bats (it would help if the names did not rhyme). Rats, like weeds and bacteria, are r-selected, producing an enormous number of offspring they take minimal care of. Bats are K-selected, so a bat mother may have only one pup per year (and "pup" is the term for it). The mother recognizes the pup's smell and voice, and can find it in a colony that may number thousands. Because they have only one pup at a time, they put a great deal of care into it, same as we do. Bats can live as much as thirty years. Rats do not do that, either.

So, they are not rats. Do not treat them as such, and at the same time, do not treat them like more friendly rodents, like flying guinea pigs. The best thing to do is regard bats as another of the park's interesting animals.

Bats are, in truth, really amazing animals. They are the only mammal that is capable of bird-like flight—that is, flight in which their wings are flapping. In this they are different from flying squirrels, which are really gliding squirrels. This killer adaptation—the biological equivalent of the engineer's killer app—is one reason there are so many different kinds of bats, living almost everywhere on the planet.

While some species of bats eat fruit, the ones in the park eat insects, using sound waves to home in on them in rather the same way a submarine uses sound waves to find other vessels without revealing its periscope (the term for the submarine's sound gear is "sonar," while for a bat, the word is "echolocation"). Because they are warm-blooded and so vigorously active, bats have to eat insects in spectacular quantities, and they also have the endearing habit of eating insects that humans despise.

Included are quite a few of those arthropod vectors we were talking about earlier: Bats can eat as many as 1,200 mosquitoes an hour. A nursing mother will eat her own body weight in bugs in a single night—in fact, she must, because of the demands nursing places on her body.

Having only one pup at a time has also made their populations vulnerable. Bats elsewhere in the country have been victimized by a fungal disease called white-nose syndrome. In spite of the name, it is not funny in the slightest: It is named for the visible white fungal growth that develops especially on their muzzles. Bats hibernate during the winter, or go into torpor, and the fungus attacks them then, moving from bat to bat through their closely packed colonies. It seems to irritate them, causing them to move around during the winter and so starve to death, having run out of fat reserves before the spring brings their insect larder back to life.

Bats were not intensely studied in the park before white-nose syndrome appeared elsewhere in the country (the fungus causing it is thought to be an exotic, perhaps from Europe). In fact, they were hardly studied at all until groundbreaking research, published in 2005, was performed that amounted to a survey of the entire Greater Yellowstone Ecosystem—the whole thing, to the extent they could. It was a timely effort: White-nose syndrome first appeared in the United States the next year. The study was run by Doug Keinath of the University of Wyoming, who wrote a summary of the project for *Yellowstone Science*. He remarked on how blank our knowledge of bats was even as recently as the early 2000s, and begins by noting the habit of bats, especially little brown bats, of getting into buildings like my shack at Old Faithful. "From such interactions," he writes, "folks have long known that little brown bats were common in Yellowstone, but the park's other bats are generally unobtrusive and shy of humans. In fact, most bats are so elusive that until recently no one really knew which species occurred in Yellowstone and nearby national parks. Experts had ideas, but no one had taken a good, hard look at the question."

The problem is the subject, which is perhaps the most elusive kind of mammal for its numbers in the area. "Except for a few colonial species that roost in large, conspicuous groups, bat roosts are often very difficult to find and even more difficult to reach. The nocturnal activity of bats makes them difficult to observe in the wild except by catching brief glimpses as they fly through lighted areas or against a moonlit sky. Also, since they spend virtually all of their active hours flying and have very keen senses, they are challenging to catch." They identified thirteen species in Greater Yellowstone (including Bighorn Canyon National Recreation Area), with appealingly goofy names like "big brown bat," "Townsend's big-eared bat," "hoary bat," "pallid bat," "small-footed myotis," and so on. During the summer, bats gather together in roosts to have their young. The scientists call that a "maternity colony," in contrast with the place where they hibernate during the winter, which they refer to using a term that sounds like a room in a medieval monastery: a "hibernaculum."

The research effort is ongoing. Inside Yellowstone National Park is the Bat Study, supported partly by the organization Yellowstone Forever, which you are likely to encounter in the park. To the east, in Wyoming, is the Absaroka Bat Census, which is focusing especially on high elevations and has found bats in large quantities up at the very roof of the continent. All this is important, because white-nose syndrome is in Wyoming now, at Devil's Tower National Monument, for instance. Scientists want to establish a baseline understanding of Greater Yellowstone bats so they will know what has changed if—when—the syndrome does appear.

Bats are certainly a reservoir host for rabies, and for dozens of other pathogens that are a problem mainly in Africa and Asia. This truth has been recognized for some time, at least since an influential paper published in 2006, a literature review paper written by Charles Calisher of Colorado State University and a team he led. If you think about it, you will see why. Bats fly; they eat insects; they are warm-blooded; they live packed together in colonies of incredible size; they have extraordinary and unusual immune systems; they do all manner of things that ensure they will be exposed to and carry microbes that make humans ill. But they don't do it much in North America. I will let Doug Keinath, a big fan of bats, come to their defense:

> *Since pet vaccination programs reduced the occurrence of rabies in dogs and cats, wild animals now represent the bulk of cases, accounting*

for more than 90% of animal rabies cases reported to the Centers for Disease Control, the majority of which are raccoons and skunks. . . . Due to an increase in negative publicity for bats, more people have started turning dead bats in to disease professionals, but reports suggest that the prevalence of rabies in the wild population of bats is small, perhaps on order of 0.5–1.0%. . . . Also, unlike larger animals, bats rarely transmit fatal rabies infections to humans. In fact, rabies from bats inhabiting buildings has been associated with only eight human deaths in United States history. The most common bat in the GRYN (the little brown bat) has never been documented as transferring rabies to humans ["GRYN" is Greater Yellowstone]. People can only get rabies from bats if an infected animal bites them and breaks the skin, and most GRYN bats are so small that it is difficult for them to break the skin. Since normal, healthy bats will usually not allow themselves to be contacted by humans (unless they are in a state of torpor during roosting), virtually all risk of exposure can be eliminated by not handling live bats. If frequent interaction with live bats is a regular occurrence, a highly effective and painless vaccine is available that further reduces risk of transmission.

If you find yourself sharing a structure with one, just keep your distance. The bat will cooperate.

Before we leave this topic, I want to do what the Beach Boys urged: I want to be true to my school. This topic is a big deal there.

I teach at the University of California, Davis, which is, among other things, in a constant battle with the ne'er-do-wells at Cornell and a couple of universities in Europe for number one veterinary medical school in the world. It's our thing. It's what we do. My family is an Aggie family: My sister is a graduate of the UC Davis vet school (she got the rabies vaccine, by the way; it's a rite of passage, when you start at the vet school, along with some other hair-raising vaccines). She met her husband when he was manager of our experimental dairy herd there. I'm an Aggie born / I'm an Aggie bred / And when I die, I'll be an Aggie dead.

One of the things that we do, as a center of veterinary medicine, is what has come to be known as One Health. The UC Davis One Health Institute explains it like this:

> *The One Health approach recognizes the growing connection between the health of animals, people, plants, and the environment. It understands that humans do not exist in isolation, but are part of the larger, total living ecosystem. The activities and conditions of each member affect others on a variety of levels: economic, cultural, physical, and more.*
>
> *Preventing disease events rather than simply reacting to them requires coordination of wildlife, environmental, human, and domestic health sectors. Prevention is always preferable to control because it actively avoids the impact of disease, and some control methods have negative social or environmental results.*
>
> *In 2008, the American Veterinary Medical Association released a report in collaboration with the American Medical Association recommending that the One Health concept be expanded across both veterinary and human health professions. The concept has continued to gain recognition since then, and future practitioners are increasingly expected to be prepared to work with colleagues across many disciplines to help solve emerging global problems.*

When I first saw that statement—I was directed to it by my pre-vet students, who were way ahead of me—I thought it idealistic. What doctor is going to look to the local veterinarians, or to the weather, as a key to clinical practice?

But that was a North American perspective warping my views. In much of the world, people still live the way they have lived for thousands of years, with the livestock walking in and out of the house, and local wildlife, like parrots or monkeys, coming and going as they please. In such an environment, when there is an epidemic—an epizootic, that is— among the animals, the human doctor is likely to want to know about it. Besides, it happens here, too. All the sorts of disease-vector and zoonotic organisms we have been looking at are the key, or one of the keys; they link the human world with the animal world, and the physical world, no matter where you go. To give one example, when hantavirus went on its rampage in 1993 in the Four Corners, it did so at the end of a wet spring. In response to the precipitation, vegetation had a banner year. In response to the vegetation, the deer mouse population did, too. The deer mice came into human habitations and brought the hantavirus with them. As a general phenomenon, it happens anew every year in Yellowstone. You do not necessarily need the National Weather Service to say whether we had a wet spring. The biting insects will tell you.

The National Park Service has gotten on board, with an Office of Public Health (OPH): "Functioning as an internal health department . . . the OPH strives for a One Health model of public health practices and integrative effort of multiple disciplines." An important part of the OPH is its Epidemiology Branch, which "applies a One Health approach to working with park visitors, concessionaires, volunteers, and partners to (1) prevent disease through education, (2) detect disease transmission if it occurs, and (3) respond to cases and outbreaks in order to mitigate disease, limit impact, and prevent future occurrences. Collaboration with internal and external partners is critical to the work of our Epidemiology Branch to help ensure parks are healthy and safe places to enjoy." Again, it sounds idealistic, and bureaucratic, but in national parks—extreme landscapes that are home to extreme creatures—people regularly come to grief by getting too intimate with the environment, and the Epidemiology Branch is good to have around when that happens, and they really do collaborate. After the Yosemite hantavirus outbreak in 2012, *Outside* magazine talked to the Park Service investigators who sorted it all out in cooperation with other agencies. Early in the outbreak, there had only been two cases, and the investigators were not sure what they were looking at:

> *"When we saw that first case, we assumed it was isolated—like all were, to that point," says Danielle Buttke, a veterinary epidemiologist who works for the National Park Service. Wildlife diseases are Buttke's specialty. She deals with tick-borne relapsing fever at Grand Canyon National Park, rabid beavers at Delaware Water Gap*

National Recreation Area, and West Nile virus in mosquitoes at Fire Island National Seashore.

Buttke got a call about Visitor One from the California Department of Public Health in July. A CDPH official told her a case of hantavirus had been reported in a California resident who'd visited Yosemite in June. Health officials couldn't be certain that she'd contracted it in Yosemite.

Then Visitor Two came down with the flu. He wasn't so lucky. The 36-year-old Alameda County man checked himself into the hospital on July 30. On July 31, he was dead.

He was the family friend mentioned earlier.

It was Buttke, with others, who found the link between the budding epidemic and the Curry Village tent cabins. They found the little mouse colonies in the walls, and they had the training to know what that meant.

We are early in the history of One Health, and it is difficult to see where it might lead. Because the problems flow in every direction, the benefits can flow that way, too. A perfect example is an affliction that affects the famous Yellowstone wolves. After they had been in Yellowstone National Park for a few years, some of them began to suffer from sarcoptic mange, or infestation with the parasitic mite *Sarcoptes scabiei*, a zoonotic species that can jump from animals to humans. That happens only rarely with wolves, but your dog can get it and can give it to you; in humans, it is the common skin condition scabies. The mite cannot complete its lifecycle on humans, so we suffer less than dogs. The mange looks, in wolves, rather like it does in dogs, with the same ugly shedding and general appearance of suffering, with that look on their faces (they seem to be thinking, "You cannot *imagine*."). It was the subject of a study by a PhD candidate from Princeton named Alexandra DeCandia, who found that what she learned about mange opened up new ways of looking at scabies in humans. Through genetic studies like her own, she said in a press release, "we can learn more about the causes of disease, the short- and long-term effects on individual animals and populations, and how best to respond as wildlife biologists and managers. For widespread diseases like mange, we can even gain insights into common processes that affect other host species—including our own—for a larger-scale 'One Health' perspective that ultimately improves both human and wildlife health."

A subject for ethicists, when a topic like One Health comes up, is the peculiar way we reserve clinical care for humans. Sending veterinarians into the countryside to treat wild animals is assumed to *not* be the ethical choice, usually because, practical considerations aside, it is thought to be an inappropriate intervention in nature's course. On the other hand—white-nose syndrome is not "natural." It is not native to North America. If there were, somehow, a way to treat it, maybe we ought to do it. It would be a fair return on all those dead mosquitoes.

Before we leave the smaller creatures for the big ones, let us have a look at my favorite Yellowstone animal, one you may be surprised to learn we have here. It will not be your favorite. It's my book, though, so here it is in a starring role: the rattlesnake.

You might guess that Yellowstone is too elevated and too cold to support such a creature, but of course, some of Yellowstone is much closer to sea level than the famous locations we associate with the place. On either side of the North Entrance, in the northern range that we visited earlier,

All but one of Yellowstone's snakes is harmless, like this one, a wandering garter snake that was zipping along and looking handsome beside Hellroaring Creek.

spreads an environment that is not the kind we associate with "Yellowstone" at all. The elevation is 5,300 feet at Gardiner. The land here is one of contrasts, where ten-thousand-foot peaks hang over the Yellowstone River itself, snaking along at half that elevation. On either side of the river is a landscape of rolling hills made of rocky debris that fell out of the glaciers when they melted twelve or thirteen thousand years ago, give or take. The word here is "arid," a place where the dominant plant is the sagebrush and a close runner-up is prickly pear cactus. Average annual precipitation is ten inches, and is regularly—often for years on end—a great deal less. It is what ecologists call a cold desert, and the rattlesnake here finds its home.

Yellowstone has a single species, the prairie rattlesnake, *Crotalus viridis viridis*. The species name comes from the Latin word for green, and it is descriptive: Compared to other species of rattlesnake, it is green in often striking hues, not Kelly green, but an olive shade that helps it blend into this low-key chromatic environment. As rattlesnakes often are, it is a large, powerful animal in its own way, with a stout body, thick in the middle, and a length that may exceed four feet, even in cold Yellowstone, where the snakes are in hibernation a large portion of the year, from October until April or May.

A prairie rattlesnake, alarmed and taking refuge in its den on a sagebrush flat below—yes—Rattlesnake Butte.

Rattlesnakes are easily recognized, even from the distance that safety requires, and which you obviously should keep. The rattle is an obvious giveaway, but the head is also striking: The rattlesnake has a marked and easily noticed neck leading to a head shaped, when looking at it from above, like the ace of spades in a deck of cards.

The same rattler, safe in its den.

The flaring on the cheeks is there to accommodate the venom glands. There is a general "rule" that venomous snakes have slit pupils, like a cat, although there are a number of exceptions—and examining pupils like a snake optometrist is a terrible way to determine which ones to avoid. Nevertheless, our rattlesnakes do have elliptical eyes, along with a pair of hollow spots between the eyes and nostrils called pits, which sense the warmth emanating from warm-blooded animals, and that create the general category we use for snakes like these: pit vipers, a name that sounds so cool it was borrowed for a brand of sunglasses.

Prairie rattlesnakes do not live in the cold desert region of Yellowstone just in general. They stick to highly specific parts—although in those places, they may be present in some numbers. They turn up especially along the courses of creeks that come down out of the mountains on either side of the Yellowstone River, including, to the north, Bear Creek, Crevice Creek, and Deckard Flats between them; to the south, they can be found at Stephens Creek, Reese Creek, and Rescue Creek. In the latter location, they are common enough that a mountain nearby, easily identified from the North Entrance and the area around it, is called Rattlesnake Butte. Watch, to the northeast, for the hill with the flat top that looks, in outline, rather like a Civil War ironclad—like CSS *Virginia*, the former USS *Merrimack*, the one that fought USS *Monitor* at the Battle of Hampton Roads (it is, naturally, a lot taller, and would never float). Prairie rattlesnakes turn up elsewhere, and in surprising places: In 2021,

A bull snake on the northern range.

I found one right next to Little Trail Creek, just west of Gardiner, but at the unusual elevation of 7,125 feet. A friend found one even higher up on the south face of Sepulcher Mountain.

In such places, the rattlesnake shares the terrain with a snake that looks very much like it: the bull snake, *Pituophis catenifer sayi*, the cause of many a near heart attack among Yellowstone visitors who stumble upon one unexpectedly. A close look reveals that the bull snake has the narrow head of a snake that lacks venom glands. It also has no rattle, and the color pattern on its back is not quite the same as the rattler's. However, they do look alike at first glance, and bull snakes have an extraordinary technique they adopt when they feel threatened that takes advantage of

Another view of a bull snake. No rattle, no problem.

this similarity, as Edward Koch and Charles Peterson explain in their book *Amphibians and Reptiles of Yellowstone and Grand Teton National Parks*. It makes them all the more frightening.

> *Bull snakes have several effective defensive mechanisms, including defensive behavior similar to the more dangerous rattlesnake. When threatened, bull snakes will often coil and flatten their heads into a more triangular shape. They also may shake the tip of the tail, similar to a rattlesnake. They often will hiss loudly before or during strikes (which are mostly bluffs). A ridge of cartilage in the trachea (windpipe) oscillates back and forth, acting as a mechanical amplifier. Presumably, the bull snake gets its name from this hissing noise, which to some people sounds like an angry bull. To some observers, it also sounds like the rattle of a rattlesnake.*

I think it sounds that way to nearly everyone, and individuals of this species can get to be six feet long and more. You can, now and then, watch as people encounter bull snakes around Mammoth Hot Springs and places nearby. They will jump and literally run, and it can be difficult to convince them afterward that the snake is harmless. I have jumped and run myself, many times.

But I did so because rattlesnakes and I have a history together. I have an unusual perspective on these creatures. It requires an explanation, and a short narrative. It will be worth your time.

A closeup of a bull snake. No venom glands, no problem.

It begins trivially and happily enough. One spring about ten years ago, I arranged and rearranged my schedule, and at the end of April, I left for the fishing expedition I had looked forward to so long, and with such longing. I had not been fishing in a year. Without getting into the ugly details, I had suffered a catastrophic financial reverse, so I went to a place I could afford, a place where we vacationed when I was a kid: Millerton Lake State Recreation Area, in the foothills above Fresno, California. As "wilderness" destinations go, it is the exact opposite of Yellowstone, the most perfect exact opposite of Yellowstone you can achieve in North America. It is an impoundment on the San Joaquin River designed to supply water to agribusiness in California's Central Valley. Most of the water goes to factory farms, to grow grapes to make fortified wine and Franzia in a box for sale at Walmart. The lake is 560 feet above sea level and blisteringly hot in the summer. The people of Fresno come here to get drugged up and drunk, race their boats, and drown.

But it is an appealing place when they are not there, and it reminds me of my childhood (I had an intensely blue-collar childhood). Because the people of Fresno go there to get drugged up and drunk and drown, the fish are, to a surprising degree, left alone.

So, I found myself this day, the first hour of a four-day weekend, on the shore of an arm of the lake that was entirely empty of humans. A nesting pair of golden eagles was riding the thermals above this arm of the lake. The vegetation was still green. A pair of bobcats had their kittens nearby the previous year, and have always favored that area. We have even seen mountain lions.

And the fish were feeding, largemouth bass. The situation could not be happier. I worked my way along the shore, climbing up and down stacks of boulders; those boulders are what the bobcats like about this area, as a place to shelter their kittens. I stepped off one and onto another, a drop of perhaps two feet.

I felt, in an instant, the strangest thing: a prick like a hypodermic needle, but more than one. And a sound, like air escaping from an air compressor. I looked down at the spot where the prick had happened. I was wearing lightweight water shoes and shorts, with no socks. A pair of dots welled up on the upper part of my left foot, bright red dots like plastic beads, a color of such intensity it seemed artificial. It was, in fact, arterial. The hissing continued.

It was replaced by a sound so loud it echoed off the opposite canyon wall: profanity, in a voice scarcely human. A bright blue streak filled the sky. My vacation, my fishing holiday, was ruined, and I let the world know it. I am sure it was audible on the far shore, a mile or more away. Happily, no one was there to hear it.

I climbed down off the rock, and there he, or she, was. The sound resolved itself into, not a hiss, but a rapid rattle, the individual beats of sound

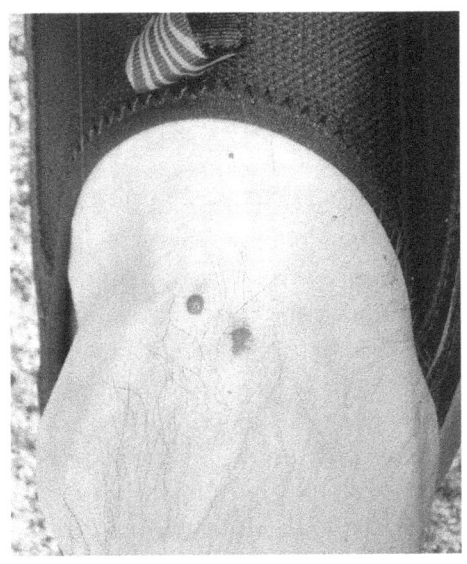

The first sign that I might be in trouble: the fang marks on my ankle, with their little beads of blood.

coming so quickly that they blended together. He, or she—we'll make her a "she"—was wedged in the space between the boulder I had stepped off and the one on which I landed. I had not known the space was there. She had backed against the far end of what for her was a cave, and had just enough room, now, to work her rattle, at great speed. The boulder itself was tall and dense enough that she no more knew I was there than I knew she was there.

Until now, as I looked her in the eyes: a Northern Pacific rattlesnake, *Crotalus oreganus*, which I was amused to learn, recently, was formerly known as *Crotalus lucifer*. It is a close relative of the rattlesnakes in Yellowstone National Park.

I had been cussing because, even though I guessed this was probably a "dry bite," to use the medical slang, I would have to head back to Fresno just to be safe. The formal name for a dry bite is a "venomous snake bite without clinical envenoming." In metabolic terms, venom is expensive.

Herpetologists (the people who study reptiles) have a theory that the animal may be able to control the venom dose. Because a great deal of energy goes into venom production, my rattlesnake did not want to waste it on a passing angler whom she was not going to be able to eat anyway. This species of snake uses venom to immobilize prey, mostly rodents—the local snakes here probably live on ground squirrels. They like rats and mice quite a bit, too.

From an evolutionary point of view, venom is an ingenious solution to a difficult problem: How do you stop your meal from running away, quick and jittery as rodents are, and dangerous as a rat can be? Venom allows even a relatively slow-moving cold-blooded animal that might weigh five or ten pounds to get the better of whatever might be on the menu, without breaking a sweat (which she couldn't—no sweat glands). People often wonder, and no, the venom does not hurt the snake when she swallows it, inside the poisoned prey animal; the venom is designed to work in the bloodstream, and the digestive system breaks it down same as it breaks down any kind of protein. It is like insulin for people with diabetes: It has to be injected to work, so the snake has evolved a naturally occurring syringe and needle.

Was this a dry bite, then? I looked at my adversary. She was not yet full grown; stretched out, she would not have been even three feet long. She had that rattlesnake look about her, though, and if you can believe it, I was not mad at her. I was mad at fate. I saw now what had happened. It was April. Millerton Lake is not natural in the slightest; the gates in the dam were set now to hold water back, and the snowpack in the Sierra was headed my way. The water level was rising every day. A few weeks ago, she was probably using the boulders that I knew, from past visits, were several hundred feet from here, and well underwater. The progressive rise of the lake level had washed her out of her home.

I still did not know how to evaluate this whole situation, and was beginning to imagine that I could just stay here and hope for the best. How would you go about discovering if this were a dry bite, except to wait and see?

And then I found out.

It was about ninety seconds after the bite that my lips and gums suddenly went numb. A wave seemed to flow over and around and through my mouth, a wave of anesthesia like the one the dentist gives you, but this wave kept going. It seemed like a palpable, physical force, a flood that surged up, reflected off the top of my head—the actual inside of my skull—and made a second trip back down to what was plainly its point of origin: the bite.

Okay. So it's not a dry bite.

What followed, though, was so strange as to be difficult to explain. The feeling, now, was not unpleasant. In fact, it was the opposite. It will be easier for you to understand if, sometime in your life, you were forced to call the paramedics for some accident and were given an injection for pain. A broken bone—that would do. If they gave you a dose of Dilaudid or morphine . . . that, believe it or not, is what it felt like. It was entirely pleasant. It was *intensely* pleasant.

Naturally, a certain silliness followed. "I suppose," I said to the snake, happily, "that that's why they call it 'intoxication.'"

The rattling continued.

"All right, all right, then! I expect I'll see you again, sweetie."

This may be the snake that bit me, and is surely a relative. It turned up in the same spot, some months later.

The snake, settling into its coil. Note the ace-of-spades head and—naturally—the rattle.

I waved. The rattling followed me as I reversed course back down the shore, but whether I saw her again, I do not know. I did eventually return to that part of the lake, with considerable caution, but left for good when I saw a rattler the size, I swear, of a boa constrictor. Maybe it was her, all grown up now.

I was down inside a canyon. This part of the lake was simply the San Joaquin River, stoppered and filling the valley it had cut through the Sierra foothills. The cell phone would not work here; in fact, I had left it in the car. I followed my footprints back the way I had come, then picked up an old jeep track that carried me over the ridgeline that makes up the wall of the river canyon. The climb was steep, through a treeless stretch of sand and boulders. I was, by the time I reached the car, pouring sweat to such a degree that it was filling my eyes and blinding me. Other than the inebriation and the sweat, there were still no symptoms.

It was hot in the car, too hot to linger, but still I sat and thought. I now had cell reception. However, recall the catastrophic financial reverse I mentioned. I remembered a long feature story I had read some time earlier in an outdoor magazine about helicopter rescues on Denali in

Alaska. The figure of $20,000 stuck in my head as the total bill for the helicopter ride.

Would they send an air ambulance? Surely, they would just send a regular ambulance, but who knew? A bill for $30,000 would break me permanently. So I turned the key and started driving.

At the entrance station kiosk, where visitors pay to get in, I found my favorite ranger on duty, Ranger Shaggy, my name for him. He looked just like Scooby Doo's friend Shaggy, and had the same happy-go-lucky style.

Should I stop and tell him what was happening? But the $40,000 air ambulance hovered again into sight: He would surely call every ranger on duty, and the fire department, and the National Guard. I slowed and waved, but kept going.

"See ya later!" he said brightly, waving.

I said, I think, "Rooby Rooby Roo!"

As I drove the long, sinuous, slow entrance road, another complication developed. As part of the financial crisis, I was making do with what turned out to be my last flip phone. I had no map program. What I did have was a Kaiser Permanente card, and Kaiser has a big medical center in Fresno. How to find it? It was somewhere in the sprawl to my left as I drove, not too fast, out of the foothills and into the urban neoplasm that is greater Fresno. I had only an imperfect idea where I might be. I talked to myself as I drove: "To Highway 41. That might work. In the part of the city . . . with all the new buildings that are . . . *bizarre*."

I looked at those buildings closely now, for the first time. One after another: private hospitals specializing in some kind of surgery, weight loss especially, or in some other chronic health problem. Weight loss, weight loss, diabetes, heart surgery, diabetes, weight loss. The area had been zoned to encourage, somehow, for-profit medicine and nothing else. It had all been grape farms when I was a kid. Gallo wine came from here. The extravagance was mocking me: health care, health care everywhere . . .

I was going to have to ask for directions in one of these places. Clinics like these are filled with the kind of doctors who no longer know what to do in an emergency. If I said what was really happening, would they not blow my cover and call the air ambulance? My $50,000 helicopter was still right overhead; granted, I was inside the city now, and a regular ambulance would respond, but that $60,000 helicopter was still buzzing right over the car roof, making clear thought difficult. I needed a fib. Like the Grinch, I came up with one, and rehearsed it: "My son, ah, fell and, ah . . . hurt himself! Where is Kaiser, please?"

The receptionists at the heart surgery place told me. Climbing back into the car, I caught a glance at my reflection in the mirrors and the glass. The top of my head was now wet as a mop, tendrils of hair lying pasted to my forehead. My face glistened, and the skin underneath was bright red.

But the ordeal was nearly through. I found Fresno Street, and after a perfect losing streak with the red lights, the familiar Kaiser Permanente logo hove into view.

I arrived just as the euphoria ended and the long hangover began. My bitten foot was swelling now, and swelled significantly for the four hours or so I was in the emergency department. A bulge was growing across the top of my bitten foot; a nurse outlined its border with a black marker, and it grew well beyond that boundary before the swelling finally ceased later that night. It hurt to stand on the foot. Walking grew more and more difficult. The foot turned blue in various shades, and at length, the leg above turned yellow.

The doctor gave me a bag of the standard antivenom, a fluid called CroFab delivered by IV drip; a couple of hours later, she gave me a second. It settled all the symptoms and even reversed them, except the swelling and pain, which took a couple of weeks. She ordered a tetanus shot, one of the unadvertised dangers of snakebite being tetanus from the deep puncture wound made by a snake that will not have been brushing her teeth. The next day, I visited my sister, the veterinarian, who lived among the cows she treated in the farmland north of Fresno. She had a good look at the wounds and saw that there were actually four tooth marks, the two deep ones and a shallower matching pair on the lower edge of the foot where the smaller fangs caught the flesh and provided the leverage by which the larger fangs could be plunged deep inside my foot. The size of the bite was astounding: The upper and lower fang marks were more than four inches apart. She was also able to explain more about what had happened.

And what *had* happened? I cannot be certain about some of it. I believe I was correct at the start: Every drug is a poison in some way, and this time, the order of equivalence was reversed and the poison became a drug. There is no clear boundary between the two categories (and hence the bartender's traditional prompt, "Name your poison"). Almost every recreational drug will kill you if you get enough of it. Lots of toxic substances create euphoria, like benzene and toluene. Believe it or not, carbon monoxide poisoning causes euphoria; a doctor can even use that symptom to reach a diagnosis. As weird as it seemed at the time, it was not weird at all.

I learned things I never expected to know. Medical websites instruct you—thank goodness—to *not* do that thing from the old cowboy movies, cutting the site of the bite open with a bowie knife and sucking the venom out. They are correct. The venom spread through every inch of my body in not much more than a minute. Cutting would have only given me another wound subject to infection, and I am certain I could not press my lips against the top of my foot tightly enough to create suction. The thought is actually kind of comical. Logically, the workout I got hiking out of the canyon should have made matters worse, by firing up my heart and circulating the venom that much more quickly, but it made no difference; again, the venom had made the whole Grand Loop circuit very quickly indeed. Nevertheless, I can see the value of a first-aid treatment invented in Australia called the pressure immobilization technique. Briefly, take the limb that was bitten and wrap the whole limb tightly with bandage material, as tightly as for a sprained ankle. No one can say for sure that it works, but it seems to help. Venom does not just move through the blood; it uses the lymph system, too, and the thought is that the pressure slows the passage of the venom out of the limb by this route.

Otherwise, what happened to me was a kind of science experiment showing what rattlesnake venom does to its victim. Venom is a witches' brew of organic and inorganic . . . *stuff*. The Arizona-Sonora Desert Museum in Tucson has, naturally, a major focus on rattlesnakes. As they explain, "Snake venoms are extremely complex; there is no 'standard' rattlesnake venom. They consist of combinations of proteins that range from hemotoxins, which break down cells and tissues, to anticoagulants and neurotoxins that may cause circulatory arrest or respiratory paralysis." Once the venom was inside my foot, it attacked the local tissue and blood vessels. Electrolytes, red blood cells, albumin, and other things spilled out of the vessels. It also attacked those red blood cells directly, tearing them apart. I ended up with a great deal of the former contents of cells and bodily structures suddenly loose and building up in the area. This resulted in the multihued discoloration and visible swelling in the foot and farther up the leg. It was a trace better, and the pain was easing a little, when my sister explained all this to me the next day.

Why do I like rattlesnakes, then, to the extent that they are my favorite Yellowstone animal? Because you cannot have an encounter like that and not come away with some respect for this creature that can upend your life like that. In part, they have the appeal for me that dinosaurs have for young boys: They are fearsome creatures of undeniable power, despite their small size. So it is that Edward Abbey, in *Desert Solitaire*, refers to the western diamondback rattlesnake, a close relative of our Yellowstone rattlers, as "the mighty diamondback." But they are also creatures of undeniable beauty, if you can get past the danger. The symmetry of the scales. The symmetry also of the coloration on the back, looking like a Roman mosaic. The grace of the movement over ground, with the snake's body formed into curves that recall the meandering of a river. The sudden coiling, and the remarkable speed of the rattle as it sings out its warning. There is much to admire.

They are dangerous, yes, but not as much as we may think. Look how my story went: It was not an experience I want to repeat, but in the end, the pain was comparable to a badly sprained ankle. I was due back at the university on Monday, and I was able to walk, if slowly, to class (the class was a writing course for pre-meds, and I talked about snakes and venom, to a rapt and only mildly horrified audience). My body's response to the antivenom was instantaneous. As is often the case with medical matters, the venom is dangerous mainly to people who are very young, very old, or in some way debilitated. Wikipedia maintains a page called "List of fatal snakebites in the United States," similar to their list of bear attacks and shark attacks. It is reasonably complete through much of the twentieth century, and only gets sketchy at earlier dates. Keeping a list like that is only possible because fatal snakebites are so rare.

The experts agree with me. Edward Koch and Charles Peterson, in *Amphibians and Reptiles of Yellowstone and Grand Teton National Parks*, argue that rattlers are dangerous, but not so much that we should lose our minds:

> *If they cannot escape, prairie rattlesnakes can defend themselves by striking and injecting venom, but they usually are reluctant to do so. In our experience with this rattlesnake species, we have been impressed with their attempts to avoid humans. During our observations of rattlesnakes at dens in Yellowstone, we never were struck at, even when we moved the snakes with snake hooks. Some rattlesnakes will actually hide their heads under their coils rather than strike.*
>
> *In Yellowstone there are only two known cases of people being bitten by rattlesnakes. . . . In 1945, John French, a trail crew laborer, was bitten and recovered quickly. And in 1886, "Red River Dick" was bitten. Only the snake died from the event.*

In spite of this (almost) exemplary record, they have not always been given the respect they should:

> *We have observed prairie rattlesnakes to be locally common. However, even in Yellowstone National Park, rattlesnakes have suffered persecution at the hands of humans. In the 1930s and 1940s, Rudolph Grimm submitted annual reports documenting the number of rattlesnakes killed by him and his staff each year. In 1938, Grimm reported to the park superintendent: "32 rattlers killed this year." Of course, at that time Grimm was simply doing what he believed was right, and what was probably his job, just as government trappers had killed the last wolves in the region a decade previously.*
>
> *Fortunately, people have slowly gained a greater appreciation for all animals, predators and prey alike, and we now are learning to accept that a rattlesnake or a wolf is not "good" or "bad," just as*

elk and deer should not be given a value based on human biases and emotions.

Think of Greater Yellowstone as a woven garment. Pull one thread, and the whole garment is marred; pull the wrong thread, and it might fall apart. This is not a new metaphor, but it is one that is useful because it is accurate. The elements of Yellowstone—water, sky, elk, bears, bacteria, viruses, and so on—are the threads. Pull one out, as has happened in the history of the region, and the damage is often greater than you could have imagined.

It is what happened with the wolves, as we will see. They were removed, violently, and it damaged the place. They were returned, and the damage was to some extent reversed. Yellowstone is resilient that way, but you don't want to damage it on purpose.

Let there be no more unraveling—which means leaving my friends with the fangs and the rattles on their rear ends alone. Besides, you can take it from me—the party is fun, but the hangover is nasty.

Chapter Five

Nature Red in Tooth and Claw

The Carnivores

We have already seen what the bears can do. What about the other carnivorous animals in Greater Yellowstone? We have some that are an obvious danger, and some others that I wonder about. We'll begin with one of the obvious ones: the mountain lion. It is a good next step from the rattlesnake, because it too was persecuted.

The mountain lion (*Puma concolor*) is also called a panther, puma, cougar, and many other names, for some reason (and it is anyone's guess why "cougar" became a slang term for an older woman who pursues younger men, in which usage it kind of looks like a compliment). The species is native to the New World, ranging from the Yukon all the way to Patagonia, and historically, before Columbus, mountain lions occupied nearly every place in between. Their range is reduced today, but they are actually not rare; they are just so secretive that they seem that way. They are large animals, but not outlandishly large. Males may measure between seven and eight feet long, females between six and seven—but this length includes the tail, which may be three feet long. Their color ranges from reddish brown to tan, all over; the "con" in the species name *concolor* means "thoroughly," because they are all one color, except the kittens, which are spotted and, with bright blue eyes, frankly adorable. According to the Mountain Lion Foundation, the male weighs between 110 and 180 pounds, the female 80 and 130. The African lion, which may weigh 500 pounds and more, is much larger, but is not a close relative. Mountain lions are actually more closely related to your housecat, and they are good at the same things Mr. Boots is, in much greater proportions. According, again, to the Mountain Lion Foundation, they can jump forty feet while running, leap fifteen feet straight up, easily climb a twelve-foot fence, and run fifty miles per hour in a sprint.

Adult mountain lions are loners. They have territories that are enormous: Males can range over a dominion measuring a hundred square miles, and they will fight to defend it from other mountain lions, if necessary to the death. They are very much carnivores, feeding on a wide range of animals but primarily deer, which they will kill at a rate of about one a week. They hunt by ambush, hiding above a path they know the deer will follow, and simply waiting. In one of his many books, George Bird Grinnell, the pioneering naturalist and conservationist, gives us a portrait of how Greater Yellowstone mountain lions lived during the nineteenth and early twentieth centuries in his book *Jack, the Young Explorer*, published in 1908. It is actually an adventure book for boys, but Grinnell uses it to record much of what he had discovered about the wildlife of the Rockies. In it, he constructs a scene in which an old mountain man is teaching young Jack about the ways of the mountain lion; the scene functions as a way of delivering information Grinnell got from Billy Hofer, a close friend. A resident of the area from the 1870s, Hofer filled for Yellowstone the role John Muir did for Yosemite, as its greatest early all-around expert. Through Grinnell's book, Hofer gives us a glimpse of mountain lion life when Yellowstone National Park was young and Grand Teton National Park did not exist:

> *"You see, the lion is at work all the time. He's got to eat every two or three days, and to eat he's got to kill something. Now and then he may pick up a bird or a rabbit or a woodchuck, but his main dependence is these animals here in the mountains. High up like this there are not so many lions, and I was surprised to see that one to-day, but lower down there are a good many, and, of course, in summer they work up higher. On the other side of the range, where deer are plenty, they kill lots of deer and a few elk, but they also kill a great many sheep and goats, most of them, perhaps, young ones.*

> "You know about their killing goats, son, for you've seen them do it, and you remember that story that I was telling you the other day about a lion jumping on what he took to be a sheep. Now, there's a place down south of here on Boulder Creek up near its head, where two men, both of whom I know well, Colonel Pickett and Billy Hofer, found eighteen or twenty skulls of sheep all by one rock. They had been killed at different times. Some of them were mighty old and all falling to pieces, and some of them were pretty fresh. They had all been killed under a high rock, not in a place where they could have been hit by a snowslide, but in a place where a lion could lie by the trail without being seen, but could himself see both ways. The rock was right over the trail, so close that a lion could jump right down on it.
>
> "The two men who found these skulls were both good mountain men and they both believe that this was a place where a lion lay and killed his food as the sheep passed along the trail under the rock."

The sheep skulls and the ambush-rock are not fiction; Hofer himself writes about them elsewhere. There are Boulder Creeks all over the western United States, but it appears this one is in what is today the Bridger-Teton National Forest, south of Yellowstone National Park (and Pickett, by the way, was an early patron of Hofer's). As we will see, a good way to avoid coming to grief with mountain lions is to avoid these ambushes they set up, the problem being that the lion is entirely hidden from view. They're good at it. Modern wildlife scientists have come to associate lions and deer, but Grinnell thought that lions in the Rocky Mountains were just as likely to go after sheep. "While, of course, it cannot overtake sheep in a fair chase," he wrote elsewhere, "the puma lies in wait for them among the rocks, killing many, because the sheep range is on ground suitable for the lions to stalk them on; that is to say, among the rocks on steep mountain sides or the edge of cañons." Yellowstone, filled with sheep, deer, elk, and other big animals, is thus a happy place for the mountain lion.

Or at least it was, until—paradoxically—Yellowstone became a national park. It was at first not "managed" at all, but once it was, the persecution began. Mountain lions were common, or as common as that kind of animal can be. A survey of the history of the species in Yellowstone was published in 2022 by researchers William Ripple, Robert Beschta, and Luke Painter at Oregon State University in Corvallis. They followed up on earlier work done by the park historian and others, who reviewed every early account they could find and learned that lions were outright common between 1835 and the 1880s. The US Army took over the park in 1886, and now the park had managers who tended to want to correct the "mistakes" they thought nature had made here . . . it seems. Actually, it is hard to imagine what they were thinking, but it led them to kill predators: "During the early decades of the park, in the late 1800s and early 1900s, there was a major government campaign to kill large mammalian predators, including cougars. Park officials at that time considered large predators to be vermin and were concerned about the welfare of their prey, mainly native ungulates. The National Park Service owned and used trained hounds to find, chase, and tree cougars, greatly aiding in their population reduction." The NPS was founded in 1916, though, and the big push happened earlier: "The main cougar culling began in 1904 and continued to 1908, during which 63 cougars were killed."

What happened in 1908 to put the brakes on? There is, in the archives in Yellowstone National Park, an extraordinary document that explains it. It is a single typewritten page, dated January 22, 1908, and addressed to the superintendent of the park, at Fort Yellowstone, Wyoming, a then-retired US Army general named S. B. M. Young. "Dear General Young," it reads. "I do not think any more cougars (mountain lions) should be killed in the park. Game is abundant." The author then provides evidence to support his position: "We want to profit by what has happened in the English preserves, where it proved to be bad for the grouse itself to kill off all the peregrine falcons and all the other birds of prey." The author concedes that lions may have to be killed around Mammoth Hot Springs if the population of grazing animals there were to fall too low, "but in the rest of the park I certainly would not kill any of them. On the contrary, they ought to be let alone." This far-sighted little document was then signed by its author—Theodore Roosevelt. Young had actually been Roosevelt's senior officer during the Spanish-American War. Now, Roosevelt gave the orders, if gentle ones, in this case.

The "last" mountain lion, on display at Mammoth.

Roosevelt was, however, only in office for another year, and Young left Yellowstone later in 1908. The killing of mountain lions continued, with another fifty-seven dying by 1919, when the killing ceased because there were no more lions to kill—apparently. A wonderful thing about mountain lions is the way they can outwit humanity. One lion was killed in 1925, and oddly, you can still see that one, because it is on display in the visitor center at Mammoth; they needed a specimen, and that last mountain lion provided it. For years, it was assumed in a lot of quarters that the mountain lion was completely extinct in Yellowstone after 1925. A good guess would be that people were confused by the disappearance of wolves from Yellowstone, which happened very close to the same date, and the wolves really were gone completely. Mountain lions, however, never were entirely eliminated; there were always a few hiding out there. Milton Skinner, the park's first naturalist, thought there were as many as a dozen in 1927, and the Oregon State researchers went through park documents and found many references to reliable sightings. Here is one, from *Nature Notes*, the in-house publication for rangers, dated July 1926: "A tourist fishing on the west bank of the Yellowstone at the site of the old Barronette bridge, hearing a slight sound behind him, turned to see a large mountain lion crouched on a rock and lashing its tail. Much frightened, the man dropped his rod and ran half a mile to the present Yellowstone bridge and down the other side to his guide, the night watchman of Camp Roosevelt." That lashing tail has a nice literary effect. The watchman saw the animal, too, so we have some confirmation that it was a lion. A half mile is a long way to run over terrain that rough; the fisherman must have been terrified to a spectacular degree.

A happy ending did eventually occur for the lions. Gradually, the surrounding states became somewhat less aggressive in their effort to drive mountain lions to extinction. Starting in the 1960s and accelerating through the 1970s and 1980s, there were more and more sightings inside Yellowstone National Park. One factor was the increasing size of the elk herd. Here is another odd and now discontinued government policy: For decades, the authorities were convinced that there were too many elk in the park, and that the elk were eating themselves out of house and home (there are people who still believe it). As a response, elk were herded together and killed in a policy that was not popular with the general public, once they found out about it. The culling ended in 1968, and the elk herd—critics were quick to point out—grew and grew. But here came the mountain lions, returning after their long near-absence; it must have looked, to those first lions, like a predator's version of Thanksgiving dinner. The only other large meat-eaters in the area were the bears, and their diet is different from that of lions. No wolves—yet. They would, as we will see, be reintroduced in 1995. Wolves kill kittens and even adult mountain lions, and when a lion makes a kill, wolves may run it off the carcass (bears will do that, too). But for years, the newcomer lions had the grazing animals very largely to themselves.

The bulk of the mountain lion population in Yellowstone National Park lives in the northern range, where conditions are ideal for them: Snowfall is rarely heavy, and there are always grazing animals to eat. Before the wolf reintroduction, between 1987 and 1993, researchers thought there were between fifteen and twenty-two lions on the northern range, counting both sexes and all age ranges. In 1998 to 2005, as the wolves established themselves after reintroduction, there were between twenty-six and forty-two. In 2014 to 2017, there were between twenty-nine and forty-five. No one is really certain why, but the wolves and the mountain lions, which are competitors, grew their populations at the same time. Of course, the elk died in quite some numbers, but it may be that their population merely assumed the size it ought always to have been. At this point in history, no one can say for sure.

Recent decades have thus seen growing populations of mountain lions. Naturally, run-ins with humans have grown in number, too. And the humans don't always come out on top.

There has never been a fatal attack, that anyone is aware of, in Yellowstone or Grand Teton National Parks—again, keeping in mind that these places have long histories that we know little about. There have certainly been attacks in the Rocky Mountains, nonfatal and fatal both, although the key factor leading to an attack is simply having a lot of people around. In the Rockies, therefore, the place to go to find lion-related mayhem is Colorado. As the *Denver Gazette* reported, in Colorado, "there have been 25 known mountain lion attacks on a human since 1990—the last verified mountain lion–related death in Colorado took place in 1997 and involved a 10-year-old in Rocky Mountain National Park. A 3-year-old may have also been killed in 1999 by a mountain lion in Colorado, though that case remains highly debated. Regardless—fatal attacks are rare and even rarer in cases where an adult human is involved." The numbers are comparable for California. There are roughly 4,000 mountain lions in Colorado, and 4,500 in California. In March 2024, there was an ugly incident in which a lion attacked two adult brothers in the Eldorado National Forest, in the hills between Sacramento and Lake Tahoe. One was injured badly, and the other killed. In a story on the attack, the *Sacramento Bee* reported that, counting the Eldorado attack, "there have been 22 documented mountain lion attacks involving 24 people in California since 1986, according to the California Department of Fish and Wildlife. The last fatal attack in Northern California was in 1994, when a woman was killed in Auburn State Recreation Area." In the whole state, it was the first fatal attack in twenty years.

The closest fatal attack to Yellowstone National Park—again, that we know of—happened in greater Missoula, so actually closer to Glacier National Park than Yellowstone. In 1989, a five-year-old boy was killed right behind his house there. It was, therefore, a more typical attack, because if a lion is going to attack a human, they prefer the human to be a child. In 2002, a mountain lion attacked an eight-year-old girl in British Columbia, and a physician, Denise McKee, wrote it up as a case report in *Wilderness and Environmental Medicine*. The girl escaped with bites and lacerations. It was, in many ways, a "typical" attack, and McKee summarized what tends to happen in such cases: "Cougar attacks on humans are unusual," she wrote, "partly because cougars do not perceive humans as prey, and cougar prey recognition is a learned behavior." That bit is crucial. Mountain lions have to learn what constitutes "food." In the millions of years before humans arrived in the New World, they never ate any and never learned to. They do it now because humans intrude on their lives so much. It may be, then, that the situation is similar to the one that develops with sharks. Humans are thin and bony compared to a shark's usual diet, and one school of thought says that when they attack us, they do so by mistake—but given what the insides of their mouths are like, one little mistake can be fatal.

McKee continued:

> *Their increasing proximity to humans . . . may have the effect of habituating cougars to humans by sight and smell, possibly identifying them as prey. Others have suggested that cougar sickness or starvation also plays a role in human attacks. Overlapping habitats and increasing cougar populations are also suggested for the increase in human-cougar interaction. Data indicate that children are at a higher risk than adults because of their small size, quick movements, excited conversation, and vigorous play movements, all of which may stimulate cougar attack behavior. It does not appear to make a difference whether children are alone or with a group of other children or adults. Cougars have attacked them under all of the listed scenarios.*

A different kind of mountain lion attack was the one at the Auburn State Recreation Area in 1994 that the *Sacramento Bee* referred to. It involved an adult—but sadly, in a scenario that would be difficult to consciously avoid.

Auburn State Recreation Area is an odd place. It was the intended location of a dam and reservoir, but the project was canceled when it turned out that the dam site was right on top of an earthquake fault. The highest bridge in the state was built before the cancellation—it had to be high to cross the reservoir—but today serves mainly as a filming location (for example, in the film *XXX*, Vin Diesel was depicted parachuting to safety after his Corvette was driven from the bridge; the car body was made of fiberglass, and people were picking up bits of it from the riverbank below for years—thanks, Vin). It was also the site of the first fatal mountain lion attack in California in nearly a century.

One of the investigators of the attack was state park ranger Jordan Fisher Smith, who wrote about it in his eccentric classic *Nature Noir*. On April 23, 1994, Barbara Barsalou Schoener, a forty-year-old Placerville resident and ultramarathoner, was running on a trail here when she literally jogged into an intensely dangerous situation she could never have guessed existed. She entered the exact situation Billy Hofer discovered a century earlier when he found that rock, in what became the Bridger-Teton National Forest, that was littered underneath with sheep skulls: The trail led her into a mountain lion's ambush zone.

In *Nature Noir*, Smith describes piecing the event together with the local sheriff-coroner's investigators. They followed the path Schoener had been running along:

> [T]he trail curved out from under the trees onto a rocky ridge surrounded by manzanita bushes that provided cover close to the path for anything that would have wanted it. One of the sheriff-coroner's deputies pointed out a divot of moss loosened from the bank uphill of the trail. As we reconstructed it, the cougar had been sitting in the brush, maybe hungry, lying in wait for the next animal to come along the path. That happened to be a runner, a woman. The animal sprang down the bank, leaving the divot as it launched. It hit Barbara Schoener from above and behind.

That would be the classic mountain lion tactic—jumping from above and using gravity as a way of adding emphasis to the blow. A deer's neck would be broken that way. Recall how closely related your housecat is to the lion, and you will see them doing it, too; a good guess would be that they like to crouch up on top of the refrigerator or the armoire because that is what their ancestors did, hiding in ambush, waiting for an unsuspecting victim. There was clear evidence of a struggle below the ambush rock, but the mountain lion persisted, and after killing Schoener, it hid her body to return later. When it did so, the authorities were there and chased the animal down.

It has been some decades—in parts of the western United States at least—since persecution by humans has slackened enough for mountain lion populations to rebound. Today, it is hard to imagine that California, for instance, ever offered a bounty on mountain lions; it did as recently as 1963. There are a lot of the creatures around now, and we have some sense of what they are likely to do. The academic journal *Human-Wildlife Interactions* (equal parts fun and creepy) published a study in 2011 that analyzed all attacks since 1890 and had some recommendations. The authors agreed that children are at greatest risk. Having a dog along on the trail seems to deter mountain lions, which is not much help inside the Greater Yellowstone national parks, but can be in the national forests. They also found that "people engaged in more erratic or high-energy activities, like running or mountain biking, were at a greater risk for attack and fatality. This may be because the prey response is more easily triggered in mountain lions from these movements." Note the use of the word "may" there; the authors were speculating on what is happening inside the mountain lion's head, and that always remains partly a guess. "Additionally, these activities may reduce a person's ability to respond quickly to a mountain lion and act aggressively, or to notice a nearby mountain lion quickly enough to stop and back away." This part of their findings fits the Schoener attack exactly.

The danger is, however, regularly overstated (I'm looking at you, Reddit). The California Department of Fish and Wildlife, which deals with these matters with great regularity, has put some energy into studying them. Researchers put radio collars on lions that travel and hunt near humans and discovered facts about them that are somehow reassuring and disquieting at the same time. The findings were summarized in the magazine *Outdoor California*:

> *The information gleaned from these collars has provided some illuminating results. They have indicated that mountain lions regularly use such areas more frequently than we have previously thought, and that these lions generally attempt to stay away from people.*
>
> *For example, in Southern California, university researchers have placed collars on these big cats in a heavily used park. They also placed trail loggers and remotely triggered cameras along popular trails to estimate human use. Surprisingly, the results indicated that some lions were mere feet away from people who were unaware of the lion's presence. During the course of this study, no reports of aggressive lion behaviors were ever reported to the researchers or park personnel.*

Something similar may happen with rattlesnakes. If you do not get close enough to activate the rattle, you will never know how close you were. So it is with any potentially dangerous animal that would prefer to remain hidden. You will never know how close you came to an encounter.

How can you keep yourself safe, then? Against the kind of ambush that killed Barbara Schoener, you can probably only stay home and shiver, which is not an appealing or even sane response. It is more likely that the lion will confront you and behave in a clearly threatening manner. It is possible—this has happened—that the lion could be rabid, in which case you want no contact at all (same as with the plague in Greater Yellowstone lions that we looked at earlier, although the danger of catching plague from a lion is mighty remote). As with any animal, you can never tell for sure what might work. A pioneering piece of research in this area was a 1991 article by Paul Beier of UC Berkeley. Among the attacks he analyzed was one that happened in British Columbia in 1970, in which a mountain lion attacked a fifty-year-old woman who was hiking alone: "As she fell," he wrote, "she set up her backpack as a shield, faced the cougar and (in her words) 'began talking to her the way you would if you were trying to soothe a dog or cat.' She continued her soothing talk for thirty minutes until she heard other hikers nearby. Then she yelled for help, an approaching hiker blew a whistle, and the cougar retreated. This strategy apparently did prevent the cougar from continuing its attack." Beier could not, however, recommend this strategy, even assuming a victim would have that presence of mind: "It was a loud shout and a loud whistle that eventually caused the cougar to retreat."

The Mountain Lion Foundation has done a fair amount of work on this matter, and has a basic list of actions you can take if you find yourself confronting a dangerous lion. I should note that these responses are not "one-size-fits-all-animals"; the recommendations for lions are different from those, especially, for grizzly bears. First, you should emphasize how big you are, and make yourself appear larger (even though you may not feel especially big at such a moment): "Make yourself appear larger by picking up children, leashing pets in, and standing close to other people. Open your jacket. Raise your arms. Wave your raised arms slowly."

You should also be loud: "Yell, shout, bang your walking stick or water bottle. Make any loud sound that cannot be confused by the lion as the sound of prey. Speak slowly and loudly to disrupt and discourage the lion's hunting instincts."

You should also project confidence and appear the opposite of frightened. Again, these recommendations are liable to run against your feelings at the moment. Do it anyway: "Maintain eye contact. Never run past or away from a mountain lion. Don't bend over or crouch down. Aggressively wave your arms, throw stones or branches, do not turn away."

You should create space between yourself and the cat: "Assess the situation. Consider whether you may be between the lion and its kittens, prey, or cache. Back away slowly to give the mountain lion a path to retreat, never turning your back. Give the lion the time and ability to get away."

Finally, if necessary, you should defend yourself, as violently as possible: "If attacked, fight back. Protect your neck and throat. People have used rocks, jackets, garden tools, tree branches, walking sticks, fanny packs, and even bare hands to turn away mountain lions." In Greater Yellowstone, it seems likely that you will be carrying bear spray—you certainly should be. Spray the mountain lion in the face. That will stop an attack dead.

Except for the spray, fighting back is the opposite of what you are told to do in a grizzly bear attack, in which struggling is regarded as useless, and playing dead is the best approach. That is not what you do with a mountain lion. Many people have fought them off, and the literature on the topic is filled with hair-raising accounts of such struggles. In 2019, Travis Kauffman from Ft. Collins, Colorado, a trail runner just like Barbara Schoener, was attacked by a mountain lion that bit and lacerated him badly. It was a small animal, not more than forty pounds, but was big enough that the struggle lasted ten minutes. The cat had its teeth sunk into his wrist and would not let go as the two spilled off the trail and down an incline. Kauffman was at last able to strangle the animal with his foot; only in death did it let go of his wrist.

But Kauffman walked away. The injuries healed, and later that year he returned to running the same trail.

A number of carnivores live in Greater Yellowstone that are dangerous in a fight, but only a marginal threat in practical terms. Of these, the scariest—but also the rarest—may be the wolverine. It does make sense that they made him the most ferocious of the X-Men.

A wolverine looks a little like a skunk, if seen from a distance, and it used to be thought that they were relatives. They are not, however; skunks belong in a family that they have, perhaps appropriately, almost to themselves. Closer up, a wolverine looks a little like a bear. It is actually a mustelid, a member of the weasel family, and a relative of otters, badgers, martens, and other animals we have in Yellowstone. The wolverine is significantly bigger than any of the others, with a large adult male weighing perhaps forty pounds, a small adult female perhaps twenty, or a little less—comparable to a medium-size dog, although you would not make the mistake of petting one of these. Their scientific name, *Gulo gulo*, comes from the Latin word for "glutton," and refers to a false belief that they take more than they need, a belief that may date to the era of the frontier, when a trapper who did not get to the animals in his trapline before the wolverine did would lose all of them, finding only shreds and patches left behind.

A wolverine has an extraordinary range of adaptations that fit it to the difficult environment in which it lives. It is circumpolar: Look at the world from the top and draw a circle around the lands that surround the North Pole, and that is their home. Most live, today, in Alaska, Canada, Scandinavia, and Russia, having been persecuted in the usual way in the southern parts of what should be their range. It is a challenging part of the world, but those adaptations make them very much at home there. Their fur is effectively waterproof. Their feet are like snowshoes. They are a good deal more all-terrain than the most wide-ranging all-terrain vehicle; they use their whole paw when they walk, and with claws that function like crampons, they can run up and down steep ice and over fallen trees without breaking a sweat (not likely to happen in their part of the world anyway). They have what in a human would be an orthodontists' delight, a molar at the back of the mouth rotated at a right angle to the direction it should be going; pointed toward the inside of the mouth, it enables the owner to tear apart meat that is frozen solid.

They have been persecuted in the warmer parts of their range, yes, but have been making a comeback in places. A wolverine tripped a camera-trap in the Sierra Nevada near Lake Tahoe, where it was not supposed to be, in 2008. They have been seen in California regularly ever since. They are not exactly thick on the ground; the whole state of Wyoming may have about 15 wolverines, and the whole Lower 48 has between 250 and 300. One place where they survive in the southern part of what should be their range is Yellowstone National Park. Much of what we know about the Yellowstone wolverines, we learned during the course of a pair of extended studies that ran during the first decade of this century. One, a multi-agency effort that looked at the eastern part of Yellowstone National Park and the nearby national forest from 2006 to 2009, found 7 wolverines, each occupying a spectacularly far-flung range in the absolute middle of nowhere. The other study, run by the Wildlife Conservation Society between 2001 and 2010 in widespread sections of terrain northwest and southwest of Yellowstone National Park, captured 38 wolverines and fitted radio collars to each. The second group arrived at an estimate, for the whole Greater Yellowstone Ecosystem, of 63 wolverines; they thought the GYE could, at maximum capacity, hold not quite 150 comfortably.

But what is a "Yellowstone wolverine"? Given the way the animals behave, the words can seem almost meaningless. Consider the saga of wolverine M56, as narrated by the Montana Department of Fish, Wildlife & Parks:

> *In 2008, researchers with the Wildlife Conservation Society captured a young male wolverine (M56) just south of Grand Teton National Park that made national headlines based on its extraordinary long-distance movements. Over the course of several months, the wolverine traveled 585 miles ending at a location in Colorado that was 336 miles, straight-line distance, from the capture location. During its movements, this animal crossed one interstate, three U.S. highways and five state highways. Movement data showed the avoidance of subdivisions. This was the first documented wolverine in Colorado since 1919.*

Was it a "Yellowstone wolverine"? It seems not to have liked the place.

There are no records of a wolverine attacking a human anywhere, so it certainly has not happened in Greater Yellowstone—again, that we know of. The reason we should include wolverines in this book is not what they are likely to do to you, but what they can do, simply because they are so awesome (and *awesome* describes all the animals in this book—a little adolescent enthusiasm can be a good thing). People who deal with them regularly, in places like Alaska or Canada, come away with a level of respect that borders on fear.

Consider the following story, which appeared on the question-and-answer website Quora. The topic of conversation was wolverines as a potential danger to people. The writer, a popular author on the site (he calls himself Only Average Stupid) recalled an incident that occurred when he was an officer in a US Army infantry battalion in Alaska. They were on maneuvers in the backwoods, in the lumpy wet ground called muskeg, when a neighboring unit was delayed. On the radio, battalion headquarters "demanded to know what the holdup was," and got a surprise:

> *There was a pause for most of a minute before Charlie 6 (Company Commander) came on the horn and said: "There is a wolverine sitting in front of the point team and he won't let us move forward." Now it was Battalion's turn to pause. After a few moments there was a terse order, "Keep me advised and catch up as soon as possible."*
>
> *[Three days later], I tracked down the XO [the executive officer— the second in command] of Charlie Company and asked him if the wolverine report was real. This is what he told me: We were moving across a muskeg flat about a mile across when we saw a dark dot running toward us from the treeline about half a mile away. When it got closer, we saw it was a wolverine. He stopped about 20 yards in front of the point team, then sat down and looked at us. As soon as the point team advanced, the thing started snarling and tearing up chunks of muskeg. The point team stopped, and the wolverine sat back down and just looked at us. The CO told them to go left, and the wolverine ran to get in front of the point team and, again, started snarling and tearing up the ground. When we stopped, he sat down and watched us. We went right and it was the same thing. Whichever way we went, he would not let us advance or go around him. When we stopped moving, he'd settle down, but if we tried to get past him he just went nuts until we stopped. Finally, after about an hour of this, he just turned around and ran back the way he came into the tree line. Never saw the animal again.*
>
> *What would have happened if Charlie had really pushed it? No idea, but no one wanted to chance it. It was clear that a 150-man rifle company did not intimidate this thing one bit.*

The stories about the strength and ferocity of individual wolverines sound like fiction, or legends, but they have been documented. They are true. In single combat, wolverines have been observed killing polar bears and moose. It is perhaps for the best that, in Yellowstone National Park, they live in remote places, far from human visitors.

We have other animals that could conceivably cause trouble for a human visitor: the fox, for instance, which we met earlier as a potential vector of rabies (its scientific name is *Vulpes vulpes*, which I will include

A fox in the wild, where they belong. GETTY IMAGES

here just to note that scientists, among themselves, call them VuVu). They have not been closely studied in our part of the world, and no one is quite sure how many there might be inside Yellowstone National Park. If you see a small canine with a reddish-yellow coat and black socks, one that looks like a fox, it is likely a fox. The animal is unusual in that it actually looks like its representation in cartoons; it is one of the only animals that Walt Disney's artists got right. Foxes are nocturnal, and so are not easy to see, but they also habituate readily when people feed them. In 1997, one took to hanging out at the Tower Fall parking area, for the food. It was relocated three times—once sixty miles away—but kept returning until finally, the authorities felt they had no choice and killed it. Something similar happened to a fox at Colter Bay in Grand Teton National Park in 2021. As the Park Service argues, and it is difficult to disagree, feeding wild animals is immoral, as understandable as the urge may be:

> *Feeding wildlife is actually a form of animal cruelty. Animals that are fed by humans learn to frequent roadsides and parking lots, dramatically increasing their chances of being run over by a careless*

motorist. Most animals have very specific natural diets and therefore specific kinds of digestive bacteria. Being fed human food causes the wrong type of bacteria to become dominant in their stomachs. Soon these animals are no longer able to digest their natural foods. They end up starving to death with stomachs full of what they should have been eating all along. What could be crueler?

Fed animals also pose a threat to humans. Feeding rodents is especially dangerous because they can transmit diseases deadly to humans, such as Bubonic Plague and Hantavirus. Simply putting yourself within flea-jumping distance (up to 10 feet or 3 meters) of a rodent puts you at risk of contracting one of these diseases. Furthermore, the majority of national park visitors who suffer rodent bites report that they weren't even offering the animal any food—they were simply extending an empty outstretched hand to lure the animal closer. But because the rodent is so accustomed to a piece of food being at the end of an outstretched hand, they often bite the hand, thinking it's food.

Foxes may bite (although, with an average size of between ten and twelve pounds, they aren't going to kill anyone—unless rabies comes calling). So might bobcats, conceivably, the smaller cousin to the mountain lion. Bobcats in Yellowstone are another enigma, the size of their population unknown. An extraordinarily handsome animal, they are as secretive as their close relative the mountain lion, and so rarely seen that the possibility of a bite is remote enough that we can actually just dismiss it completely. Don't try to pick one up—but a bobcat would never allow you to get that close anyway.

A badger can bite, too. It is another handsome animal, with striking alternating white and dark stripes running front to rear on its face, and a pair of round ears like a bear's. They spend a good deal of time underground; they are adapted to digging, so much so that in Mexican Spanish, the word for "badger" is *tlalcoyote*, Aztec for "ground coyote." They are also nocturnal. However, if your stay in Yellowstone National Park includes some time on the northern range, you stand a reasonable chance of spotting one. If you do, try to get a photograph. Yellowstone's tracking ace, Jim Halfpenny, is currently running a research project powered almost entirely by citizen science. Halfpenny, a longtime resident of Gardiner, has become the Gandalf the Grey of tracking; he is a legend locally for his ability to follow animals by their pawprints alone, and such other traces they might leave behind, often subtle in the extreme. Halfpenny's website for his badger project is at www.yellowstonebadger.info. If you spot a badger and get a good photograph, upload it there. As the website explains, "The charismatic megafauna such as bears and wolves draw the attention of Park visitors, managers, and researchers. Less is known about smaller carnivores because of their low population densities, sparse spacing, secretive habits, and difficulty making observations." He launched the project to correct this problem.

The badger is an aggressive creature. It is a predator, hunting small animals, especially the Uinta ground squirrels present seemingly everywhere on the northern range. It can make unearthly noises, including a shriek that leaves the animal issuing it sounding dangerous indeed. The name alone sounds aggressive. But the word "badger" in the sense of "harass" comes from the old blood sport of badger baiting—setting dogs on badgers in a fight to the death. It ultimately comes from the same word root as "badge," and refers to the white stripe on the animal's face, more obvious on the European badger species. I suspect that if you badgered one sufficiently, the animal would rip you to shreds. You would deserve it.

This kind of listing is all improbable and speculative—if you see a fox in the park, for instance, you will probably be catching a glimpse of the animal as it dodges from one hiding place to another, and it is not likely to approach and bite. But you can just never tell. Yellowstone is the land of the improbable. A good way to demonstrate what I mean is by taking a closer look at a Yellowstone animal that you probably will see during a visit, and that ought to be too small to be a threat.

The coyote is not quite small, but contrasted with its cousin, the wolf, it does look that way. For some years, the National Park Service has used, as a visual illustration of size differences, an artist's rendering of a fox, a coyote, and a wolf drawn side by side. Compared to the coyote, and especially the fox, the wolf is Godzilla. But that may be a reflection on the terrific size of the wolf more than the smallness of the others. At perhaps thirty pounds and standing not quite two feet tall at the shoulder, coyotes are comparable to the family dog, and look like one, even though coyotes also have a strong coyote-ness about them that most people recognize almost instinctively. The two are close relatives: Your German shepherd is *Canis familiaris*, the coyote *Canis latrans*. As with the "ground coyote," their name comes from Nahuatl, the language of the Aztecs.

Coyotes eat almost anything that comes their way, although certain themes emerge, and much of what we know about coyotes has emerged specifically from Yellowstone National Park. Groundbreaking work was done here in the 1930s by Adolph Murie, a Park Service wildlife biologist famous today for his work on wolves. His 1940 book *Ecology of the Coyote in the Yellowstone* was wildly controversial inside the NPS, and elsewhere, because it took the side of a predator. Government policy of that time, of course, was simple: All predators should simply be killed. Murie had spent time on the ground getting to know the local coyotes in intimate detail. He tells us in the book that, in order to determine their exact diet, he analyzed 5,086 coyote scats (he calls them "droppings"), finding "8,969 individual food items." Given that these were coyotes he was picking up after, his research strategy required that he include a whole page of "Miscellaneous Food and Nonfood Items." These included "Horse manure . . . Garbage . . . Corn (refuse) . . . Paper . . . Canvas-leather glove . . . Rag . . . Butter wrapper . . . Twine . . . Banana peel . . . Orange peel . . . Leather (1 piece containing rivet) . . . Cellophane . . . Steak bone . . . Grape seeds . . . Mouse nest material . . . 7 inches of curtain . . . Pear . . . Prune seed . . . Match . . . 2 sq. inches rubber . . . Tinfoil . . . Shoestring . . . Mud . . . Paint-covered rag . . . 8 inches of rope . . . 3 sq. inches towel . . . Lemon rind . . . Bacon rind . . . Two pieces of shirt . . . Canvas . . . Gunny sack," and so on. It is not clear how he was able to identify some of this stuff, but his devotion is impressive.

Coyotes are not, however, dependent on human waste the way some animals can be. In Yellowstone, there is usually some form of animal protein available, or insect protein; in the 1990s, a Yellowstone friend of mine worked with a famous coyote researcher who told him that coyotes "will eat grasshoppers like popcorn." You can verify that yourself, if you are up to it. Look for canine droppings, which can appear anywhere in

Yellowstone, except maybe the hotel room. You know what the stuff looks like, and while it might be from another canine, a good bet is that it came out of a coyote, given the way they can turn up almost anywhere in the park. If that animal has been eating grasshoppers, the abdomens are hard to miss, even without having to gather the stuff up and pick through it the way Murie did (and you may see fur there, too, which will confirm that this is a wild animal dropping; dog food doesn't come with fur). During the winter, coyotes—which do not hibernate—will eat a great deal of carrion, so that a hard winter is actually good news for them. Murie identified a partnership with ravens, which find a deceased elk, bison, deer, or other large animal quickly. When the coyotes see the birds circling, they will head that way (and so did Murie—while he was there, it was a three-way partnership).

Coyotes live in packs, as do their close relatives, wolves (which also know to watch the ravens). Because they live in groups, they exhibit an intelligence that is inevitable for a creature that has to remember which individual animal is in charge, which is not, which is getting old, which is the strongest, which is the weakest, which might be alpha one day, and so on. Since the roles and relationships are changing constantly, instinct alone will not suffice; the rapid changes make an unchanging instinct unworkable. It is a general rule of human life that we feel the greatest affection for animals like these, because they are like us; we, too, live in groups and have to remember which individual is in charge, which is not, and so on (we certainly speculate about who might one day be the alpha, the animal in charge). Some of the affection we feel for our dogs is transferred rather easily to the coyote, although it is never as intense.

And in fact, you might not like coyotes at all, since they have become a threat to the nation's pets. Coyotes are unusual in that, in an era of history when so many animals have been persecuted, they have actually expanded their range, and dramatically. When Europeans discovered the New World, coyotes were confined to the prairie and desert areas of what is now the United States and Mexico. Today, there are a great many more wild coyotes in the world than there were in 1492. They are absent only from Hawaii; otherwise, they are in every US state. They live along the edges, and sometimes right in the middle, of enormous cities. They like it there because, as we have seen, they will eat anything, and in the city, the predators that at one time kept their numbers down are absent. David Quammen once traveled to Burbank, California, center of the entertainment industry, to examine the urban coyote population there for *Outside* magazine. "What sort of coyote," he asks himself, "has Los Angeles created?":

> *It's a creature that will jump over chainlink for a bowl of Alpo. It's an animal that can learn and remember which storm-sewer channels lead to which golf courses, which duck ponds and swimming pools offer potable water when the hills are dry, which dumpsters behind which supermarkets are likely to be overflowing with old vegetables and delightfully rancid fish. It's a beast constantly on the alert for unattended barbecued chicken. It's a predator that, like some two-legged ones, is at home on Mulholland Drive. It has eaten from the Tree of Forbidden Knowledge, and it recalls fondly the taste of Fifi and Mr. Boots.*

They are every bit as adaptable as humans. Many find that endearing. Many pet owners, those understandably enraged when the coyote eats Fifi or Mr. Boots, not so much.

Intriguing questions arise about what has happened to coyotes during this period of expansion, and the Yellowstone coyotes have made a contribution toward answering them. In the 1990s, Robert Crabtree completed a lengthy study of coyotes, focusing, during the later years of the study, on the coyotes of Yellowstone National Park; he was the researcher who told my friend that coyotes eat grasshoppers like popcorn. The Yellowstone Park animals were and are an "unexploited" population: They have never, that is, been subjected to the kind of shooting, trapping, and poisoning that has taken place elsewhere in North America, where the rancher carries a rifle in the pickup partly with the coyote in mind. Quammen spoke to Crabtree while the study was still underway, and before the reintroduction of wolves to the park, which lowered the population of coyotes substantially. Crabtree was reaching conclusions about what the centuries-long slaughter of coyotes had done to the animals

This is one of the five-by-eight-foot sheds Bob Crabtree's researchers used during his coyote project; the antenna on the roof picked up signals from the radio collars worn by coyotes on the northern range.

themselves (a slaughter which Quammen, by the way, found appalling, and he did not bother to hide it). The "ADC" Crabtree refers to here is Animal Damage Control, a federal agency that is today called Wildlife Services. A part of the Department of Agriculture, its mission has been to kill animals that farmers and ranchers find a nuisance:

> *Sixty years' worth of largely indiscriminate killing by ADC trappers and others has inevitably reshaped the American coyote, producing an animal that's more clever and wary and resourceful, more problematic, than ever before. The fittest have survived, and doggone if the fittest aren't harder to trap, harder to poison, harder to fence out, harder to fool, harder to kill despite all the helicopters and leg-hold traps and high-powered rifles and cyanide booby traps that ADC can muster.*

> "They've created their own worst nightmare," says Crabtree, not without sympathy for the many trappers and ranchers he has gotten to know. "They've created a coyote that's impervious to their means."

The main thing the long assault on coyotes has proven is that coyotes are indestructible. Short of nuclear attack, they cannot be entirely destroyed.

Are they dangerous to people? Sometimes, yes, despite their small size and not especially aggressive nature, although as with many of the life-forms we have looked at in this book, they are much more dangerous to pets. By the year 2006, there had been 160 attacks in the United States, according to an article that ran in *Smithsonian* that year (one with a clever title: "City Slinkers"). As with several other animals we have looked at, Wikipedia maintains a site that catalogs all attacks—and as with the others, the fact that the attacks fit on a single page confirms how rare they are. There have been two fatalities. In 1981, a three-year-old was killed in Glendale, California. Most attacks have taken place in greater Los Angeles, of which Glendale is a part. The second fatality happened in a national park in Nova Scotia, Canada: In 2009, a nineteen-year-old woman was killed in Cape Breton Highlands National Park.

The attack on the child in Glendale was the work of a single animal, but the Nova Scotia case involved at least two. The whole affair was so odd that a group of wildlife biologists went to work trying to learn what had happened, and they got answers: The coyotes in this park had come to specialize in using the pack to kill moose. Coyotes are, as noted, not large. Moose are, as we will see, quite large. The coyotes came up with this adaptation because Cape Breton Highlands is nothing like Los Angeles; food is so hard to come by that extreme measures were needed. "The lines of evidence suggest that this was a resource-poor area with really extreme environments that forced these very adaptable animals to expand their behavior," Stan Gehrt, a wildlife ecologist at Ohio State who was involved in the study, told the university newspaper. "We're describing these animals expanding their niche to basically rely on moose. And we're also taking a step forward and saying it's not just scavenging that they were doing, but they were actually killing moose when they could. It's hard for them to do that, but because they had very little if anything else to eat, that was their prey," he said. "And that leads to conflicts with people that you wouldn't normally see." After they had taught themselves to kill very big moose, killing a somewhat big human looked easier than it might have, and so they went ahead and did it.

Coyotes are, however, not nearly as dangerous as dogs; between thirty and fifty people die in the United States every year from dog attacks. For that matter, a human bite is a far more common injury than a coyote bite. Indeed, at some point in medical school, you read up on human bites; as many as 150,000 occur every year. If treatment is delayed, infection and amputation become a significant risk. The commonest large animal in Yellowstone is the human, anyway (and let us also note that "Biting" carries, in the National Hockey League, a five-minute major penalty, and a "Game Misconduct").

I would be inclined to dismiss this topic, and skip the coyote entirely. Still (as noted earlier), a forty-three-year-old woman who was skiing alone in January 2020 on the Grand Loop Road, near South Rim Drive at the Grand Canyon of the Yellowstone, was attacked by a coyote, and suffered puncture wounds and lacerations—it was not just a single bite. Canyon Village is a busy place in the summer, but there were only about twenty employees living in the area at that time, as is normal in the depths of winter; the victim was one. Another, Jeff Henry, told a reporter from the *Billings Gazette*, Brett French, that snowcoach drivers in the area had seen a coyote acting aggressively. It may be that the animal was not in its right mind. Doug Smith, a park biologist, told journalists, "We suspect this coyote may have been starving due to having porcupine quills in its lower jaw and inside its mouth. Its young age likely led to its poor condition and irregular behavior." Rangers managed to kill it, and answered the one question at the forefront of everyone's mind at moments like this. No, the animal was not rabid.

The *Billings Gazette* regards itself as Yellowstone's newspaper, and so did a more thorough job with the story than anyone else. "Some unusual coyote attacks on humans have occurred in Yellowstone in the past," French wrote, "although the park has no exact number because of inconsistent record keeping":

In January 1990, a 27-year-old Illinois man and park employee was cross-country skiing in Biscuit Basin when a coyote lying along the trail leapt up and bit him on the face. He suffered cuts and puncture wounds on his head, face, neck and hands, but managed to deter the coyote's attack by beating it with one of his skis. Three coyotes were later killed by park officials.

Perhaps the most frightening coyote attack in Yellowstone occurred in November 1960 at Mammoth Hot Springs, when a 1½-year-old baby who had been left in a stroller on a porch in the headquarters compound was attacked by a coyote.

A neighbor saw the assault and swatted at the coyote with a broom. The child received 21 stitches to her face and had bruises on her back and arms, but was otherwise protected by a heavy snowsuit.

For strange coyote incidents in Yellowstone, a German tourist being bitten in October 1992 stands out. The 65-year-old man pulled his car over along Sylvan Pass after the coyote was seen biting at his car's tires, according to a news report of the incident. When the man stepped out of the auto, the coyote jumped in and began eating food on the car seat. The tourist was bitten as he shoved the coyote out of the vehicle.

So, we should not skip the coyote entirely.

I also note, with interest, that nearly all these incidents happened in the winter, or close to it. They nudged my memory. As with the rattlesnake, I have an odd relationship with coyotes. We have a history together, and again, I have to explain. We are looking for an honest alternative to the fake life, and these creatures provide it. They also might help answer a question you may have: You may be wondering what the devil the appeal of Yellowstone in winter could possibly be. Here it is.

On January 6, 2012, I was in the middle of a week spent in the park. The goal of the trip was to finish teaching myself cross-country skiing, a process that had been slowly coming along for a period of years. I was alone on the Chittenden Loop, a ski trail that follows one of the original park roads, climbing from the campground at Tower Fall to join the Grand Loop Road near the present Buffalo Paddock picnic area. Almost none of this part of the Chittenden trail fails to run uphill. Once on the Grand Loop, I planned on heading toward Dunraven Pass until the sun started to sink and it came time for the long, relaxing speed run back to Roosevelt. I had now reached a point in my skiing history at which I no longer fell forward, headlong, during this speed run. (I used a GPS to measure the speed once—eight mph.)

Just as the road went flat, at the point on the winter ski map where the trail turns from "More Difficult" to "Easy," a coyote tumbled out of the forest up ahead, down from the ridgeline to my left and onto the trail. It was a hundred yards away when I spotted it. I had seen so few animals of any sort on this trip that I was immediately pleased. Coyotes have, to me, always been rather like the bison and the elk, but with quirkier personalities: I have never found them a dramatic sight—we have crowds of them where I grew up in Los Angeles County, for goodness' sake, and the smaller ones always look so weedy—but again, they do have personalities.

I recalled my closest approach ever to a coyote, one I surprised in Death Valley during my park-beatnik days in the early 1990s. I was in a shallow gulch, hidden by creosote bushes, when one passed me twenty or thirty feet away, and entirely unaware of my existence. Rarely do we see a wild animal commit such a total screw-up. I said, "Hey," and, startled, the coyote rocketed away until, at half a football field's distance, it paused to have a look at me.

This one on the Chittenden Loop turned its head toward me immediately. And then things got . . . surprising.

The animal looked at me, and without a pause, it ran right at me. If it had not been so small, I would have said it was charging.

A wild animal was coming at me, running as hard as it could. I did take an adrenaline hit. I happened to have my camera out and turned on. I tucked it under my left arm. Time slowed down. I pulled the Velcro on my right glove, tried to yank it free, then remembered to unwind the ski pole strap. With ski poles on, your hands are almost literally tied. I got the pole loose, tossed it against the other pole, and yanked at the coat pocket zipper. The bear spray was inside the pocket. The last bear sighting I knew of had occurred at Old Faithful on New Year's Day, and they should have all been in their dens, but the spray has more than one use. In the summer, I carry bear spray the way Wild Bill Hickok carried his gun, ready to fire in one second, but the threats had receded enough that it now stayed buried deep in that pocket. I yanked twice on the zipper, the first try a clumsy one. The animal was closing in, and I could not fathom this behavior. I had never seen anything like it in the wild, only in films of animal attacks, actual attacks. The second pull on the zipper got it down. I withdrew the spray, got the safety clip loose with one try, and aimed directly at the coyote, which had finally slowed, fifteen feet from me. I would not have won an Old West gunfight, but I did get the stuff out in time, and I did, by some buried instinct, stand my ground.

They are not large animals, as I have said, but I remembered the attack on the skier that January in 1990, back when I worked at Old Faithful. I remembered the sign that stood beneath Specimen Ridge asking hikers to watch out for the local coyotes, which had been behaving aggressively toward people. I was not quite scared, but found this behavior perplexing. It seemed aggressive, like a bluff charge from a grizzly. A coyote should not be big enough to pull that off, but who knew what the coyote thought? At least my first and worst thought—rabies—was dwindling. This animal was healthy and surprisingly nonchalant.

The close encounter was not over. I was of two minds: Either it meant to frighten me off, or it had learned that skiers are normally Sierra Club members, and so harmless, and some of the less enlightened ones will produce a bologna sandwich for the coyote. I had to dismiss that theory, however: If an animal wants a bologna sandwich, it begs, like the old roadside black bears from before 1970. Extortion works for the Mob, but not here—unless, a more remote chance still, this animal had learned the trick perfected by the "mugger bears" of Yosemite, which charged every hiker who was carrying a backpack. At one time, I have been told, the official advice given to Yosemite hikers stated that they should, on encountering an aggressive bear, drop their backpacks and flee. The backpacks were supposed to work as a distraction. The bears knew only that if you charged humans, they dropped their lunch.

At the same time, I was safe. I had my thumb on the bear spray trigger, and having fired that stuff for practice, I knew it would blow the coyote away. It might not even survive. I was therefore strongly disinclined to fire. This was the first time I had ever had to face down an animal so close with pepper spray out, the safety clip off, my thumb on the trigger, and the muzzle aimed at a target that I would have hit squarely, if I had fired at that instant.

But again, I was of two minds. I was safe. So I remembered my camera and thought about recording this little drama.

The close encounter was not over. The coyote got closer still, running up the ridgeline and closing within ten feet. Why up the ridgeline? For the safety and intimidation of high ground? It came even with me, above my head, then turned and ran back to a position a few feet farther back from where it had first paused in front of me. Then, I think, it decided that I was not going to be frightened off, and that I was not going to surrender a bologna sandwich, either.

The coyote moved off, a short distance, then farther. I skied without using my poles, dragging them along. The coyote would move, then stand sideways on the trail, looking the other way, giving me a quick glance as I approached within forty feet, sometimes thirty, several times twenty, then moving on. I kept skiing without my poles. I finally slipped and toppled sideways.

It hurt. "This is all your doing," I said to the coyote. "You probably think this is funny." The camera had landed on my body, but my bear spray hand landed in the snow. I made sure the muzzle was clear, then decided I could risk putting it in its holster—safety clip still off, because I was still not sure what was going on.

I strapped my ski poles back on, and we began a lengthy pas de deux. I would ski forward. The coyote would stand, looking in the opposite direction the whole time, giving me one look as I got within forty feet, then opening the distance to more than a hundred.

And now, at last, I began to think—both sensibly and creatively, for a change.

The coyote and I had been through about a half dozen repetitions of this routine: stop/look away/let me approach/look at me/move. In the end, we did it twenty times or more. I looked down at the pepper spray, still ready to fire. The muzzle, by chance, was presently pointed at my face, the safety clip still off, even though I had now firmly decided that the coyote was merely playing a highly eccentric game with me. When I had fallen, I had come as close to pepper spraying myself as I hope ever to come. I put the safety clip back on. Then I looked hard at the coyote, close again, but ignoring me—except that it could hear every shush of the skis.

Then the coin dropped. Think of how the media and various self-appointed experts have driven a wedge between us and nearly all wild animals. Few people in an advanced industrialized country are free of the comprehensive anxiety that has resulted. Think how fearful a city-dweller is of everything wild. The purpose of this book is not to add to the anxiety, but to communicate some of the awe I feel in the presence of these creatures, and that you can feel, too. Some people, of course, will take this wrong, and the anxiety will grow still more. We will revisit this issue in the final chapter. At present, my point is just that every individual fear is a brick in the wall, between ourselves and Yellowstone.

This coyote meant me no harm. I was now forming an entirely new theory about its behavior (I wish I could identify the sex, but they hold those bushy tails back between their hind legs, and it is impossible for an English professor to tell). There were coyote scats and urine scent marks all down the Chittenden Road. Skiing in this area the previous few days, I had found, on the Grand Loop, even more of the same. I had decided—and still think I am correct—that all day, I was running up and down seams between coyote pack territories. Or do they mark within their territory, overdoing things a little, as intelligent creatures will do? Every intelligent creature has the potential to become Clark Griswold, the father in the *National Lampoon Vacation* movies. But all day, I was also wishing Adolph Murie and Bob Crabtree were here with me.

Perhaps, when this coyote first emerged from the trees, it had been like the coyote that emerged from behind the creosote in Death Valley long ago—that is, it had not known I was there. But I was on the pack's boundary line, or was in a sensitive area, like the old border between East and West Germany, or the armistice line between the two Koreas. Maybe that explained the scent marking. Seeing me here, in violation of the pack's space, it had first, and instinctively, tried to drive me off. With so much scent marking—as with the constant saber-rattling between the two Koreas—perhaps things are tense around here. When I refused to be driven off, it stood its own ground, then moved back to rethink the situation. I followed it at roughly its own pace. Then it decided to tolerate me, rather the way tribes like the Cheyenne, in the old days, would tolerate a member of another tribe if he came for a visit and paid all the courtesies and respected the social rules. As we went along, the relationship eased into regularity. I was not part of the pack. I had not gone through the dominance hierarchy routine, and I think coyotes are too smart to mistake a six-foot-tall humanoid covered in various kinds of sophisticated plastic for another coyote. But I was Dian Fossey, or better, Jane Goodall, the woman who was the first to integrate herself into a chimpanzee troop. I was being allowed to accompany the animal on a boundary patrol.

Then I did something of which I am still a little proud. I put my money where my mouth was. For one thing, if it really came to that, my ski poles were perfectly adequate weapons to defend myself against an animal of that size. I had been reading, for the fun of it, histories of the Roman Empire before I left for Yellowstone. A legionary soldier would have agreed with me that the tips of the poles would serve perfectly well as an emergency defensive weapon. But then I went further. Again, we have been convinced by our culture that all animals are dangerous. Bull. (But bulls, as we will see, are genuinely dangerous.) I was on patrol with this coyote. We were companions. That word worked its way into my thinking, especially when I remembered that the veterinarians at my university call pets "companion animals." But this was no pet. This was a companion. And I should have the courage of my convictions, now, because I knew that the conclusions I had reached were the truth. So I took the bear spray, stuffed it deep into my coat pocket, zipped it shut, and quickly forgot it was there.

Our walk together went on and on. According to the GPS I was carrying, it went on for four miles. Again, I would approach to within, eventually, less than twenty feet. The coyote would give me a look—merely a glance, a confirmation that I was still along for the ride—and move ahead eighty or a hundred feet. When we reached the Grand Loop, it continued, although now, the behavior changed. At that point, a skier leaves the trees and, if headed upward, has a better and better view of the enormous open area north of Mount Washburn, visible to your right if you are driving from Canyon to Tower. Given better vision and a wider field to examine, the animal started pausing to look across the valley (but not to smell, or at least not with penetration—the wind ripping down off Dunraven Pass had robbed its wonderful canine nose of part of its power). It let me get closer and closer before it moved on.

I remembered the warnings about harassing animals during the winter, really a matter of common sense: During this time of the year, when calories are hard to come by, forcing an animal to move can ultimately be fatal for the animal. The warning fits grazers more than this coyote, for whom the time of carrion and easy living was at hand. But was I harassing this animal? I thought hard, and decided I was not. This was my companion. I had intended to head up the road anyway, and planned to go as far as I could, if the wind the night before had pushed snow over the road. It had not, but the snow was deep enough to protect the skis. Besides, the coyote had long since taken charge of our relationship. I deferred to its judgment.

We were together for a long time, long enough for leisurely thought, warped a little by fatigue from exertion and altitude; we were, by this point, approaching 7,600 feet. It was cold. The sun on the snow, now that I was out of the forest, was penetrating. So (I was thinking), this animal is my companion. I cannot communicate with it in human language, even though I have spoken to it repeatedly throughout this day, saying things like "Quit looking at me like that," or occasionally yipping in something closer to coyote language. It paid no attention to either. Then I stopped myself from doing it, intentionally. What might I be missing? I couldn't ask questions, but was there maybe some other way to talk?

What can you learn from a coyote? Plenty. Although this whole experience went on for hours, I only had occasional breaks for speculative thought. Here, however, is one example of a behavior I might learn from: the way the coyote now developed the habit of stopping, over and over and over, and looking at the open valley below the Grand Loop Road. A modern person will normally pay no attention to a detail like that, but a premodern person who has to find all his calories scattered around a wilderness might find it useful—a premodern person like the Shoshone, Crow, and other tribes that lived here originally. The coyote, so much more skilled at reading the landscape than most people, might notice something worth having, like a dying elk, a mile or two down there.

The more I thought about it—shushing along and watching—the more I understood that the coyote knew much more than I did. We call them dumb animals, but they are slick at what they do, same as we are. This one was, I think, patrolling the pack boundary, but was also looking for whatever might turn up. At one point, it found a couple of little bites to eat on the road. I would not have seen that. But I had food at home, and those bites were probably pretty gross. The only marginally edible thing I saw for days were scats, and one dead mouse.

But now the affair got a little more complicated still. By chance, I looked behind me. A second coyote was strolling along forty or fifty feet back there. It had been following me, I do not know for how long. The second animal sauntered past me with hardly a glance.

The second coyote joins us.

It approached the first coyote, which put its tail between its legs and dropped its head. The two, clearly members of the same pack, performed what I guessed was a dominance hierarchy session, one that I did not entirely understand. The second animal was a trace larger—an inch taller maybe, a pound or two heavier. They chased each other's tails in a tight circle, smelling each other intimately as canines will. Then the second animal fell in, and the two of us became three, climbing the Dunraven road and patrolling the boundary.

The second animal found my presence so unremarkable that it scarcely ever looked at me. My path through the snow was now determined by the snow itself. The wind had scoured the road and, in this light-snow winter we were having, had gone so far as to reveal the asphalt in places, even the yellow stripe down the middle of the road. I weaved, looking for snow deep enough to protect the skis. I would have kept back if I had the option, but following the snow brought me within ten feet of the second animal. It showed me what it thought of me. It found a patch of asphalt, warmed by the sun, asphalt that I now saw was part of one of those turnouts that are always filled with twenty cars in the summer. At the moment, the nearest human was at Tower Fall, a thousand feet below and miles away, if anyone happened to be visiting the fall just then. Otherwise, we were alone. The second coyote plopped down on the asphalt, yawned, and looked to be nodding off.

They showed their indifference at all times.

The point, on the Dunraven road, at which we finally turned around.

I at last turned around at 7,600 feet, more from tiredness than anything. The first animal was now far ahead. The second had fallen into the habit of moving ahead, plopping down, resting with its nose into the wind, and letting me catch up, at which point it repeated the process. It had passed me almost close enough to touch, and repeatedly, but otherwise never interacted with me at all. It was as if I were not there.

I was as baffled by their actions as I had been all along. "When I skied up there last, there was wolf sign all along that road, all along, and plenty of it," one park expert said to me, when I told the story the next day. She had the same thought I had: "Maybe they're just feeling very territorial. That first animal ran at you because, like people do, he was saying, 'Who do you think you are? I'm going to show you who's in charge.'" In a complex way, that was what the whole long walk up toward Dunraven Pass was about. They were letting me along on their patrol, but they were going to keep an eye on me and let me know that I was doing it at their forbearance.

By now, though, I was thinking of them as friends. As I turned to point the skis downhill, I looked at the second coyote, the wind blowing its hair, eyes closed from both the sunlight and the wind. It appeared to be dropping off at last. I pushed off with the ski poles—and we had another plot twist.

Through all those hours, both coyotes had been completely silent. Now, I heard a loud and strikingly coyote-like "Yip!" I turned. The first animal was running about as fast as it could, as fast as it did when it first ran at me. It had to turn to get around me, but otherwise treated me only as an obstacle, because we had an emergency. I could now see a third coyote, at the exact spot where I had turned around. It threw a long black shadow, an intruder in the snow, and clearly belonging to another pack. The first animal had run right into the third, and had immediately decided to exercise the better part of valor.

The two stopped next to me to stare the third one down. It stared right back. At some point during this afternoon, I had slipped mentally into the pack I was with. It was my pack now, and I saw the third one as an outsider, bad, not to be trusted or tolerated. This shift in attitude probably explains my fear, soon discarded as silly, that this was a wolf. A wolf, from that angle, and given my mood, would have looked as big as a moose, and likely would not have let that first animal get away so easily.

But I wonder if the fellow-feeling was mutual. When the two stopped to stare down the stranger, they did it right next to me. Was I now a force-multiplier? Did the big, clumsy humanoid, covered in bright, ridiculous plastic, now make them more formidable to the stranger? The third animal did stop and came no farther. It had reached a border it would not cross.

The second coyote showed its attitude in the same way it had with me. It plopped on its belly and, still staring down the stranger, yawned. No, it did not display its teeth aggressively. It just yawned, too bored, it seemed, to be concerned. I am afraid it looked rather dog-like as it lay there, as if it should have a newspaper in its jaws. But that too was an effect of the aloof attitude.

As is normal in such situations—in the world of wild animals, if not with humans—the border skirmish ended without combat. The two stared a minute or so longer, then turned to follow me. We continued in the same manner as before, one in front, one in back, each stopping periodically, all three of us comfortable with each other. Then, at long last, they began to weave, exploring the terrain on the downhill side of the road, reappearing for a stretch, then exploring some more. They then took one more thoroughly coyote-like action: They vanished. Somewhere past the top of the Chittenden Loop, they disappeared like ghosts. I have no idea where or how. They did, at least, leave me thinking.

Normally, encountering any wild phenomenon, like a coyote pack, I would look the animals over and quiz myself, checking to see how much scientific knowledge I had retained about the creatures from whenever I had read about them last, or last listened to an expert. This is a common way of interacting with the outdoors, a process like running a set of mental flash cards through the mind. It can be annoying, because life in the outdoors becomes just like school. You are continually studying for a test, even though there will be none.

For a long while after this incident, it was different. I did not ask, What does science know about the coyote? Instead, I would ask, What does no one know about the coyote? At the very least, which of its habits are surprising or unexpected, and what can it teach me?

They taught me how to patrol a family or pack border: Move eighty or a hundred feet, then have a good hard look. They taught me that you shouldn't worry about things that won't harm you. That when an unusual and harmless but kind of amusing human shows up, you should take him along for the ride. He may prove useful: He might scare up a prey animal, or scare off an enemy, frightened by the skis or the clinking of the poles. That was one thing that seemed to bother the coyotes, or that at least they took notice of—the clink of a pole tip on rock, ice, or asphalt.

They certainly taught me confidence. I had arrived at the correct evaluation. There are "experts" who will say I was wrong, that wild animals are all unpredictably terrifying. I would, before that day, have had doubts. Now I know, in this case as in others—many others, one of their lessons— that I am right and the "experts" are wrong. Here, too, was a major lesson: When you patrol a boundary, do it with confidence. Do not worry about things that do not rate worry, and focus on what is immediately important. I had serious problems to deal with at work, but all day long, I forgot the troubles completely.

Up there, I did feel good, and at the same time isolated. When the coyotes left, the nearest human was still at Tower Fall. Not up there in the sun, on the ice, in the wind, with Mount Washburn hanging above it all.

When this incident happened originally, I was asked by some prominent Yellowstone wildlife experts to relate the full story. I told them part of it just after the incident happened, but they wanted the rest of the details. Several of the people who heard it, I noticed, gradually let their mouths drop open. So I wrote the full story you just read for the website Yellowstone Reports, owned and operated by wolf expert Nathan Varley. The website is very much alive; if you want to know about the latest doings of the wolves, that is the place to go. This story is no longer online, though, so I rewrote it a bit and put it here, as an illustration of just how odd things can get with these creatures.

As noted, they can be dangerous. The Urban Coyote Research Project is a team of researchers who are capturing and following Chicago coyotes—and yes, there are a great many in greater Chicago, about two thousand. One of the leaders is Stan Gehrt, the Ohio State wildlife ecologist who helped investigate the fatal attack in Nova Scotia. They have recommendations about how to avoid trouble with coyotes, which they have boiled down to six steps: (1) Do not feed them; (2) do not let pets run around loose; (3) if you encounter one, do not run away; (4) consider repellents or fencing, which they have found to be effective in some situations; (5) do nothing that might make the situation worse, like hazing a coyote that is minding its own business; and (6) report aggressive coyotes right away.

I will add a seventh: Keep in mind that these characters are capable of anything.

The canines have gotten bigger as this chapter has progressed. We will finish with one that is large indeed.

The gray wolf, *Canis lupus*, is in a strange category: It is both a recent arrival and an animal that has been here the better part of forever, and certainly since the last of the local glaciers melted. They are another animal—

like mountain lions and coyotes—that were hated for decades by just about everyone, including government employees who shot, trapped, poisoned, and just generally persecuted the wolf until it disappeared almost entirely from the western United States. One of the places from which the wolf vanished was Yellowstone National Park; the last park wolves were killed in 1926. There were sightings during the decades that followed, including a spate between 1968 and 1975. I knew an elderly ranger who spotted a big, black animal in his headlight beams during this period. He was entirely certain it was a wolf. The story he told had an eerie authenticity about it, and he was an unusually qualified observer, so I think it likely that he either saw a wolf, or some kind of demonic apparition. There was, however, no reproducing population in the park or anywhere close.

After passage of the Endangered Species Act in 1973 and the addition of wolves to its list of such species, the law required action. It was postponed for years, for a simple reason that is not often stated: Ronald Reagan and the elder George Bush occupied the White House for twelve years starting in 1981, and the Republican ethic of the time kept the wolves out. The election of Bill Clinton changed everything. The solution, it was clear, was reintroduction: Wolves would have to be captured someplace where their population was still large—Alberta, Canada, was the ultimate choice—then released in Greater Yellowstone. Clinton's election led directly to the well-known photograph, taken on January 12, 1995, of his Secretary of the Interior, Bruce Babbitt, and three other federal officials carrying a box full of Canadian wolf to its new home in the park. The wolf must have been perplexed, but when released, it knew what to do.

That wolf was one of a group of fourteen, in two shipments. It and the others were not turned loose immediately. No one could say what these animals would do; wolves are amazing travelers, and for all anyone knew, these might take off and run back to the Canadian wilderness where they were captured. They were therefore placed in special "acclimation pens" in the park. When the pens were opened, they did not exactly stay put. Wolves began almost immediately to radiate out and repopulate the western United States. They are all over now. Descendants of those pioneers are, for instance, living in California; I will also venture a prediction, and say they will turn up in Yosemite in the not-too-distant future.

The magic moment: The wolves arrive, January 12, 1995.

Here is another prediction: When that happens, people there will freak out. The new wolves of Yellowstone tore into the park's population of grazing animals, the ones that scientists and park managers had long thought to be too great in number. They might or might not have been, and the question is now moot, because these wolves will never allow their population to get too big again. They found the pickings especially rich in the park's northern range, and there a new social phenomenon—another total surprise—developed. In that open terrain, people could see the wolves and follow all the drama of their lives, even to the extent of identifying and keeping track of individual animals. Some found it utterly addicting, and you will see these people in large groups at turnouts in the northern range. What happens is that the local wolves will turn up, often at a location they have frequented before. The news will go out electronically, and the wolf fans will converge. You will see them especially on the roads that radiate away from Tower Junction; look for tens of thousands of dollars' worth of high-quality spotting scopes on tripods, all pointed in the same direction. In Yellowstone jargon, these people are the wolf watchers, and they have accumulated an impressive mass of information about their idols.

The wolf watchers, though not an especially large group of them.

Those idols are, as noted, big. The larger wolves weigh 130 pounds, and they can get bigger still. They live in packs that can number as few as two and average just under a dozen (10.8 at the end of 2023, according to the Yellowstone Wolf Project Annual Report for that year). The world record for largest pack is held by Yellowstone's Druid pack, which numbered thirty-seven at its largest. Packs are named for a prominent feature of their territory, in this case Druid Peak, often with a plural "s" when the name is used in conversation (as in "The Druids have done it again"). Because they live in packs, when one turns up, a bunch will often follow. Yellowstone packs tend to be larger than elsewhere, and it is thought that the massive population of grazing animals has driven the number higher.

You can get a sense of the size of these different animals from their tracks. This is a footprint left by a coyote in the snow. The ski pole basket is four inches across.

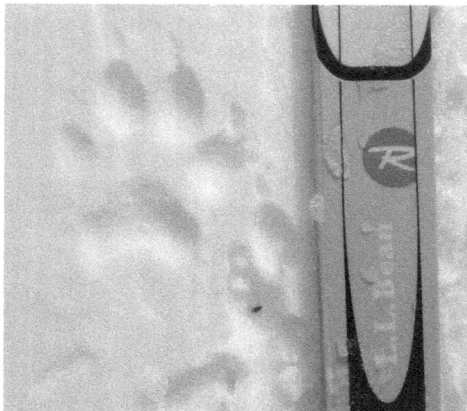

This, by contrast, is a track from a wolf. The ski is not quite four inches across.

A pack is led by a single dominant pair, described using the first letter of the Greek alphabet: an alpha female and an alpha male (when a forceful man is described as an "alpha male," wildlife biology is the ultimate source of the term). They mate for life, and they are the only animals in the pack who reproduce. They normally have a single litter of pups—the correct term—in the course of a year, in a den typically dug into the side of a hill and near water. A famous den, to the extent that a hole in the ground can be famous, is near Slough Creek Campground. The pups are born in the spring, and the rest of the pack brings them food until the pups are seven or eight months old, at which point they start traveling with the pack.

The new stars—the wolves.

That the Slough Creek den is famous is another confirmation, if we needed one, of the impact the wolves have had on every aspect of life around Greater Yellowstone. They are the new stars of the show. People write poetry addressed to them. Photographs of wolves adorn millions of walls. There are wolf screensavers, wolf refrigerator magnets, wolf potholders, wolf salt and pepper shakers. Could these much adored creatures be . . . dangerous?

Yes. I have written about this matter elsewhere (in my book *Strange Yellowstone*), and since no one has gotten too angry, I will hit the high points again here. We have not had an incident in Yellowstone—yet, but consider the following.

In 2009, there was an incident that many locals had considered inevitable. A young male wolf started hanging out in the busy part of the Firehole geyser field. Biologists, and others, keep track of the Yellowstone wolves by giving them a sequential number followed by "F" or "M" for the animal's sex, and you can thus tell, in a piece of writing about a wolf, where you are in the history of the reintroduction by the size of the number. This one was 729M of the Gibbon Meadows pack, a yearling. He turned up at the Midway Geyser Basin in March of that year, and as the spring progressed, was seen more and more regularly around Biscuit Basin and Old Faithful. He chased two-wheeled vehicles, hung out around the employee housing, and at last began to regularly approach people at Biscuit Basin. He also approached cars. He seemed to be panhandling. The rangers watched him closely for two weeks. They chased him away from public areas, using noisemakers and shooting at him with bean bag rounds and rubber bullets. He kept coming back and coming back. To the extent that we can ever know what is happening inside an animal's head, this one appeared to be habituated, just as happens with bears: 729M had become accustomed to people, and had come to associate people with food. When they felt they had no choice, the rangers shot him.

It happened again in 2011. Wolf 812M belonged to the Mollie's pack. As noted, most packs are named for a geographic landmark. An exception is the Mollie's, named for Mollie Beattie, the head of the US Fish and Wildlife Service during the reintroduction (she is in the famous photo with Bruce Babbitt, carrying one of the first wolves; she is on the left side of the image, with Babbitt on the right). Beattie died young of brain cancer in 1996, and the pack was named in her honor. The Mollie named 812M had also come to associate people with food. Seven times, they hazed him away. They even hit him with paint balls and sprayed him with bear spray, which ought to have soured him on humanity forever, but he, too, kept coming back. The wolf likely doomed himself when an extraordinary incident happened, the record of which is still on YouTube. On September 20, 2011, a man walked away from the East Entrance Road, inside the park, and turned to find a wolf, 812M, behind him. A bystander filmed as the wolf circled the man and approached over and over. The man can be heard on the video yelling at the top of his lungs, swinging with a heavy stick he turned up on the ground. "Ha!" he positively bellows, on the audio, "Ha!" over and over again, like a muleskinner. The wolf still would not back off, and kept circling the man as if trying to get behind him. Bizarrely, a turn signal indicator can be heard through an open car door, ticking like a time bomb. And 812M was a big wolf; we know that, and it is not just an estimate. The rangers eventually killed him, too, and during the necropsy found that he weighed 110 pounds.

After both of these incidents, the press releases talked about visitor safety—that is, explaining that the animals were removed to keep the public from harm. A habituated wolf has come to associate people with food. The belief is that eventually the wolf will give in and kill the goose that laid the golden bologna sandwich. It will come to look upon people as food.

In North America, it has not happened very often—but it has happened, and appears to be happening more often, an inevitable outcome now that we have decided to tolerate wolves. Not long ago, the best thinkers among environmentalists agreed that wolves did not attack people, or did so only when deranged by rabies, or only so rarely as to make them about as dangerous as meteors, which do hit people, but rarely enough to make such occasions fit only for Ripley's Believe It or Not! The writer Barry Lopez produced, in 1978, one of the classic books on wolves, *Of Wolves and Men*. Calling the book classic, however, is not the same as calling it infallible. Lopez is dismissive of the idea that wolves could hunt people:

In a book called Adventures in Error, *Vilhjalmer Stefansson recalls his efforts to track down virtually every report of a wolf killing a human being between 1923 and 1936. Reports from the Caucasus, the Near East, Canada, and Alaska all proved to be either fiction or gross exaggeration. Furthermore, Stefansson could not substantiate a single report of wolves traveling in packs larger than about thirty. In 1945 it was reported that no incident of wolf attack brought to the attention of the US Fish and Wildlife Service in the preceding twenty-five years could be substantiated.*

This passage is representative of a certain kind of environmental thinking that developed during the era after all the wolves had been killed, when there were no attacks because there were no wolves. Lopez and the authors he cited were wrong, but we have more information now, because we have more wolves.

And we also simply have more information. Once again, Wikipedia maintains—and I love them for it—a page called "List of wolf attacks in North America." The editors (working for free, of course) included every story of every attack they could find, including some tolerably flaky ones, which have the advantage of being fun. One such is the death of Caroline Allen, eaten by wolves in late eighteenth-century Vermont, date not known (and sadly, the Wikipedia gang took it off the list sometime during the last couple of months—it was too much to swallow, as it were). The best source is John Greenleaf Whittier's 1831 volume *Legends of New England*, and a couple of others from a hundred years after the event. Caroline was one of a group of youths returning from a "quilting frolic" on a deep, dark winter's night. "Did you ever see a wild wolf," one of the survivors asks Whittier rhetorically, many years after the terrible night—"a fierce, half-starved ranger of the wintry forest—howling and hurrying over the barren snow, and actually mad with hunger? There is no one of God's creatures which has such a frightful, fiendish look, as this animal. It has the form as well as the spirit of a demon."

The kids are in the dark, deep woods when the wolves set upon them. They climb an oak tree, and from their branches can watch their tormentors: "We could distinctly see the gaunt, attenuated bodies of the wolves beneath us, and every now and then we could see great, glowing eyes." In the pitch black, they saw this. Naturally, one of the branches gives way, and Caroline Allen tumbles to her horrid doom:

> *"A light form went plunging down through the naked branches, and fell with a dull and heavy sound upon the stiff snow."*
> "Oh God! I am gone!"
> *"It was the voice of Caroline Allen. The poor girl never spoke again! There was a horrible dizziness and confusion in my brain, and I spoke not—and I stirred not—for the whole was at that time like an ugly, unreal dream. I only remember that there were cries and shudderings around me—perhaps I joined with them—and that there were smothered groans and dreadful howls underneath. It was all over in a moment. Poor Caroline! She was literally eaten alive. The wolves had a frightful feast, and they became raving mad with the taste of blood."*

The others survive. Her boyfriend is present at the scene, and, haunted by the ghastly tableau, goes on to drink himself to death.

Such is the punishment meted out to unchaperoned youths who abandon themselves to debauchery and attend a . . . quilting frolic.

What this is, however, is a familiar thing, if you know it. It is the kind of tale ministers told their congregations, in a stricter era, to warn that the wages of sin, and quilting, is death. We rightly read something like this and dismiss it. It also sounds just like a romantic novel of that era (and Whittier, after all, calls the story a "Legend"). It could have happened, but I suspect it didn't.

History is filled with stories like this, and they can lead us to dismiss the reality—especially now, when we are so distant from the experience of our rural past, and there are still relatively few wolves in the Lower 48. Think of all the wolves that show up in the stories our forebears told each other, like "Little Red Riding Hood," which exists in sanitized and unsanitized versions. In the story as modern children tend to get it, the wolf locks grandmother in a closet and is chased off by the woodsman before he can do anything to Little Red Riding Hood. In older, more stern versions, he eats grandmother and the girl, and in some of those older tales, that is the end of the story; after the meal, he falls happily asleep.

When you start looking for big, bad wolves, they turn up everywhere. In "The Three Little Pigs," the first two pigs are killed by their foolish disregard for the inevitable arrival of the wolf; in "The Wolf and the Seven Young Goats," a tale collected by Jacob and Wilhelm Grimm, the wolf pretends to be the goats' mother and, having gained access to the house, eats all but one of them; in "The Wolf and the Fox," another tale collected by the Grimm brothers (who traveled Germany saving tales from wherever they could find them), a nasty, gluttonous, greedy wolf lives with a fox and forces the fox to do all the work under threat of death; and in "Peter and the Wolf," a Soviet Russian tale, the young hero is warned by his grandfather that the wolf will eat him if he plays outside alone. Or consider the ancient tale "The Boy Who Cried Wolf," which, suffice it to say, does not end happily. In some versions, the wolf only eats the flock, leaving the boy alive to suffer; in others, the wolf eats the boy, too.

More serious, "adult" myths and legends are much the same. There are friendly wolves, like the she-wolf that suckled Romulus and Remus, the mythical founders of Rome. There are, however, far more fiends, like Cailleach, the evil old hag of Gaelic folktales, who brings winter and death and comes riding on a wolf. And you definitely do not want to run into a werewolf.

These stories were the work, and the common inheritance, of people who lived close to the land, keeping animals that were *always* vulnerable—morning, noon, and night—especially at night. We have forgotten what a catastrophe it was to lose a pig or sheep, just one, and wolves could go through the whole herd. And they could eat the shepherd, too.

Incidents of wolves preying on humans have been, in North America, as rare as attacks by any of the other creatures we have looked at. Captive wolves occasionally attack, and keeping one as a pet guarantees daily interaction between wolf and human, which can go terribly awry—in a sense, it is habituation taken to an extreme. Wolves can also have rabies, but even deaths from rabid wolves are rare. Wolves sometimes "attack" defensively, or as an act of aggression, unrelated to hunger. If you meet a wolf, do not provoke it. You laugh, but people do these things, particularly when alcohol is involved (which is, for instance, a major cause of rattlesnake bites: young, drunk guys fooling with snakes). What we are concerned with here are predatory attacks, those in which the wolf, or the whole pack, is hungry and has picked you as a meal.

Even in North America, where attacks have historically been vanishingly rare, it has happened. In 2010, snowmobilers near the town of Chignik Lake, Alaska, came upon the body of Candice Berner, a thirty-two-year-old teacher who had been out jogging when she was attacked by wolves who very much intended to eat her. Because it was March, the details of the event were written in the snow. Berner was a petite woman, under five feet tall, and her size may have led the wolves to choose her over some other victim. The Alaska Department of Fish and Game investigated and found signs of a protracted struggle. Multiple sets of wolf tracks appeared 250 feet from the place where they presume she was killed. She was running, and so were the animals; it appeared to the troopers that the animals had to run to catch up with her. They found a mitten two hundred feet away from the death site, and another mitten sixty-three feet away. One of the mittens had a thumb torn off, so the assumption had to be that there was a literal running fight.

In the winter, wolf activity lies written in the snow. This was the Carnelian Creek pack.

Forty feet from the spot where she was killed, the troopers found a depression in the snow with blood in it, with the wolf tracks converging on the depression, so it was here that they first knocked her down and injured her seriously. The same thing happened again ten feet farther on, but from here, the evidence of the tracks changed. Berner had been running along a road, but now she left it and moved downhill; she had been struggling, and then crawled away. Then the tracks changed again, and it was clear that she was no longer crawling. Instead, the wolves were pulling her downhill, away from the road. There was a great deal of blood here, so she was now severely injured. The wolves pulled her to a point where the troopers found a wide patch of melted snow with blood in it. They decided that this spot was where she died. The wolves dragged her body at least twice afterward, but there were no further signs of struggle.

The Alaska DFG troopers killed wolves in the area and used DNA to tie the killing to at least one of them, and possibly more. From the tracks found at the scene, it was clear that more than one wolf was involved. For some reason, the pack had simply decided to treat Berner as they would a moose or a caribou.

A similar incident happened in 2005, when twenty-two-year-old Kenton Carnegie was attacked, killed, and partly eaten at the mining camp of Points North Landing, Saskatchewan, and this incident is especially important to people concerned with the Yellowstone wolves. The wolves in the Carnegie case were habituated; they were eating garbage from the camp, and had approached people in the area. Ten months before Carnegie's death, a lone wolf attacked one of the miners. Then, on November 8, 2005, Carnegie left to go for a walk and did not come back. When searchers found his body, it was surrounded by wolf tracks; as the body was being recovered, witnesses saw two sets of eyes glowing from reflected light in the woods nearby, and howling echoed in the distance. Carnegie had been partly eaten: The pathologist, after the autopsy, reported that 25 to 30 percent of his body mass was gone, from his midsection to his thigh. As with Candice Berner, this was a case of outright predation. Some suspicion remains that the culprit may have been a bear. Fatal attacks by bears are a good deal more common than fatal attacks by wolves. However, the tracks argue otherwise.

There are more records of attacks earlier in history, but those records are often sketchy, regularly not even naming the victim. The Berner and Carnegie attacks are the only recent fatal incidents involving healthy, wild wolves actually hunting people with the intent of eating them.

I have been speaking, however, of North America. The Old World is a different matter. There, thousands and thousands of incidents have occurred, and in these incidents, the wolves were hunting. They killed their human victims because humans were an item in their regular diet.

An important source in this matter is a book-length report by John Linnell, a wildlife biologist at the Norwegian Institute for Nature Research. Linnell led a team of seventeen other researchers in a massive effort to comb historical records and get at the truth. Their report, "The Fear of Wolves," was published in Norway in 2002. They carefully excluded cases that were plainly fictional, and those that were unreliable, they marked as such. Still, they found plenty of real cases. Most were cases of rabies—but then there were the other stories, true stories in which the wolves did not have the excuse that they were possessed by a virus.

In a single county in Poland, in the single year of 1819, nineteen people were recorded as having been killed by wolves. In an episode remembered in history as "the Wolf of Gysinge," during a three-month period in 1820–21, a single wolf in Gysinge, Sweden, attacked 31 people. In Kivennapa, Finland, between 1839 and 1850, what was thought to be a single wolf killed one adult and twenty children. Also in Finland, in a series of incidents remembered as "the Wolves of Turku," between 1879 and 1882, possibly as many as thirty-five children were killed by a single pair of wolves. Perhaps the most famous series of wolf attacks in history occurred in France in the eighteenth century. It is remembered as *La Bête du Gévaudan*, the Beast of Gévaudan, and in spite of the singular noun, it was a pack of wolves that had developed a taste for human flesh. Between June 1764 and June 1767, there were as many as 210 attacks there, with perhaps 113 killed, maybe fewer, maybe more, and a number of them were eaten. The victims, as so often, were mostly women and children; it was they who, like Little Bo Peep, were the traditional shepherds in the countryside.

Linnell recently returned to this subject (with coauthors Ekaterina Kovtun and Ive Rouart) with *Wolf Attacks on Humans: An Update for 2002–2020*. Just as wolves have been returning to the United States, their population has surged dramatically in Europe, and in some remarkably improbable places:

> Since the low point in the mid-20th century Europe's wolf populations have dramatically expanded to reoccupy large parts of the continent.... This includes returning to areas from which they were totally exterminated like Finland, Norway and Sweden in the north, Germany, Denmark, the Netherlands in the west, and across the entire Alpine arc of northern Italy, France, Switzerland, Austria and Slovenia in the centre of the continent. They have also expanded their ranges in other countries like Spain, Portugal, and peninsular Italy where they had been reduced to small fragments. There are currently (i.e., 2016) an estimated 17,000 wolves in Europe, not counting Russia, Ukraine and Belarus. Viewed from the lens of wildlife conservation this represents one of the great success stories of the last 50 years.

As recently as the 1970s and 1980s, commentators confidently predicted that the wolf would go entirely extinct, or at least would be driven to the outermost margins. Today, we have them in the Netherlands, and in California. The question now is, When will their population stop expanding? In what other outlandish places will they turn up?

Which will lead to trouble. Having so many wolves around means that trouble is not entirely avoidable.

Linnell and his coauthors took on literally the whole world: They looked for wolf attack cases in every nation where wolves live. They found so many that we will have to focus just on North America. In 2013, a sixteen-year-old boy, camping at a lake in Minnesota, was bitten on his head by a wolf that was, in this case, not at all normal; it was not rabid, but was literally brain damaged from an injury earlier in life that had only imperfectly healed. In 2019, in Banff National Park, Alberta, there occurred an incident rather like a predatory bear attack: "A wolf tried to force entry into a tent on a campground at night with a family sleeping inside. While the father of the family tried to scare the wolf away, it bit him multiple times on the hands and arms, dragging him from the tent. Neighbouring campers assisted and by kicking the wolf and throwing stones at it were able to get the family to safety, although the wolf remained nearby, following them as they retreated to a car." Rangers—"park wardens," in Canada—were able to shoot the wolf later that day, and DNA tests confirmed that it was the correct wolf. In this case, as in other cases of attacks by wolves, the animal proved to be in poor general condition, and was driven to this unusual behavior, it seems, by hunger. Such was the case in another incident in Canada, on Anderson Islands, British Columbia, in 2007. There, a wolf attacked a kayaker, who was actually able to stab the wolf and so drive it off.

Worse than hunger and poor body condition, perhaps, is habituation. When wolves become accustomed to being around humans and their towns, and to either begging or stealing human food, the situation becomes dangerous quickly. In part, it is purely mathematical: The animals are around humans all the time, and so are more likely to have some kind of run-in, especially since humans keep dogs, and wolves do not like canine competitors, indeed will kill them on sight if they feel so moved. Few breeds of dog are comparable in size to a wolf, and no dog is bred to live in packs. They are far more vulnerable than humans are.

Since the late 1990s, coastal British Columbia has become a case study in what we will need to avoid in Yellowstone. Trouble between wolves and outdoor enthusiasts has become outright routine. Vargas Island has been a hot spot. In the beginning, there was an attack by a wolf on a kayaker camping with a party of eighteen others. The victim, a man who slept apart from the rest, in the open, awoke to find a wolf dragging his sleeping bag, with him still in it. Alarmed, he shouted, and the wolf attacked. He ended up with lacerations on his back, hands, and scalp. He had, in fact, essentially been scalped, if incompletely; it took fifty stitches (not a round number—the exact count) to reattach his scalp to his head. That was in July 2000, and the wolf pack involved was thoroughly habituated; they had gone so far as to stroll right up to earlier parties on the island and play with them, as if they were dogs. That is one way that wolves get into trouble, and it is not their fault. They so often look just like Siberian huskies, and people respond to them based on that appearance—but they are not dogs. Not hardly.

Linnell reports on the work of a researcher who looked at Pacific Rim National Park Reserve, British Columbia, and the nearby area over a five-year period during which the trouble with habituated wolves happened all the time:

> *From 1999 to 2003 he summarises 51 close interactions between wolves and people and/or their dogs. The cases range from wolves entering tents, playing with or stealing campsite equipment, growling at people, approaching or following people, taking food handouts, approaching dogs, and even killing dogs. Several of the closest interactions were documented on Vargas Island in the months before the July 2000 attack, which remains the only episode where a person was hurt. Media searches reveal that the situation with habituated wolves displaying bold behaviour has continued, with dogs being attacked, even when on a leash.*

Kayakers carry items like food in a part of the kayak called the hatch, which gives access to internal space that is perfectly secure under a hatch cover. It didn't work on Vargas Island. In a feat worthy of a bear, the wolves there taught themselves to take the hatch cover right off.

In 2020, in Port Edward, British Columbia, a man was knocked over by a wolf who wrapped his teeth around the man's leg and tore out enough muscle mass to require three weeks in a hospital. The wolf was incredibly bold, circling the first responders as they rendered aid, and returning to the area later. Again, the problem was habituation:

> *In the subsequent days Conservation Officers investigated the circumstances, and it became apparent that there had been many sightings and incidents involving severely habituated wolves in the Port Edward/Prince Rupert area in the preceding months. Several wolves had been frequenting the towns' landfill garbage disposal site (c. 4 km away) where staff had claimed that the wolves fed on food even during daytime and showed no fear of humans or of vehicles. Although the landfill site was fenced with bear-proof electric fencing, the wolves were able to crawl under the lower wires. Wolves had also*

been involved in attacks on dogs, both on and off the leash, and were also apparently attracted to a significant population of stray/feral domestic cats in the Port Edward neighbourhood.

Denali National Park developed problems with habituation, too. The habituated wolves were not as evil-minded and vicious as elsewhere, but they were trending that way:

Although located in Alaska, Denali receives over 600,000 visitor days per year, mainly during summer. Road access is generally very limited and mainly closed to private vehicles, such that most visitors spend time in the back-country. During the period 2000–2007 over 250 events were logged where wolves displayed behaviour that was viewed as being habituated or bold. In most cases this consisted of curious approaches or failure to run away, although there were cases of damage to camping equipment in 7 of the 8 years. None of these episodes involved injury to people, although people had to aggressively frighten the wolves away on multiple occasions. Most events were associated with a single pack that tended to den close to the park's only road and near two campsites.

Above, I used the words "evil-minded" and "vicious" for a reason. I was hoping it would anger a few people. Friends of wolves—and there are a great many in the world, probably more friends than enemies—do not like to hear it suggested that there is anything wrong with *Canis lupus*. I used those words so that you could experience the shock. You may never have heard a contemporary author speak in such abusive terms about wolves. You may in the future, because that is exactly how wolves were described (as Big and Bad, remember) for all of human history until roughly the middle of the twentieth century. In order to stop it, in the future we will have to deal with rogue wolves as they develop.

Because develop they surely shall. Pretending that they do not have it in them to kill people is just that: pretending. Fantasy. And it is dangerous for the wolves.

We do not have to go back to killing all wolves again, but we do need to understand that they are not Labrador retrievers. Don't feed them, obviously, but also don't be so careless with food or garbage that they find it. If the wolves that killed Kenton Carnegie at that mining camp in Saskatchewan had not been feeding on the camp garbage, would he ever have met them? Would they even have been in the area? And did the proximity of humans in their everyday life teach the wolves that they need not fear these clumsy, bipedal creatures who run so very slowly?

So let them remain wild animals way off in the distance (the Park Service rule is that the distance should be at least one hundred yards). Having read the information above, you may no longer be eager to walk up to one anyway.

Chapter Six

"Man Is in the Forest!"

The Deer Family
In the summer of 1959, Yellowstone made its mark on literary history, not for the first or last time. Driving a 1953 Chevy sedan they had borrowed from a relative, Ted Hughes, future poet laureate of Great Britain, and his wife, Sylvia Plath, drove through the park. They arrived during the height of the postwar travel boom, and during the height, also, of the roadside bear extravaganza, when park visitors were allowed to feed the black bears that obligingly parked themselves all along the roads, expecting handouts.

Plath is probably best known for her novel *The Bell Jar*, and for ending her own life in 1963, after separating from Hughes. She wrote about the Yellowstone visit in a short story she published in 1961, "The Fifty-Ninth Bear." Many years later, Hughes wrote a long poem with the same name, telling the story of that same trip from his point of view. Neither the short story nor the long poem tells an especially elaborate tale. In both, a young couple, both persons too intelligent for their own good, tours the Eisenhower-era park and gets on each other's nerves. Plath's story has a fun conclusion, in which—spoiler alert—the wife apparently uses witchcraft to command one of the park's bears to kill her husband, and the bear duly mauls the husband to death.

Of interest here is a moment in Plath's short story in which the couple, Sadie and Norton, are driving on the Grand Loop and Sadie, at the wheel, spots a cluster of elk. Norton, half-awake in the passenger seat, has summoned them, literally: He imagines that he has a mystical power to summon animals if he concentrates hard enough (although in the end, it is Sadie who has the real—homicidal—power over the wildlife):

> "Elk!" Sadie exclaimed, like a voice out of the depths of his head. The car swerved to a halt at the side of the road. Norton came to with a start. Other cars were pulling up beside them and behind them. Timorous as Sadie was, she had no fear of animals. She had a way with them. Norton had come upon her once, feeding a wild stag blueberries out of her hand, a stag whose hooves could, in one blow, have dashed her to the ground. The danger simply never occurred to her.

There follows the kind of scene available in the park today. The traffic suddenly halts. People charge out of their cars, cameras at the ready, running off the pavement and into the roadside meadow, literally chewing up the turf in their hurry. The elk, disturbed by the commotion, rise and move off into a nearby forest, to the disgruntlement of both the tourists and Sadie, who is herself after a photograph.

Norton is protective of Sadie, unusually so; part of her irritation with him seems to arise from this protectiveness. When he had "come upon her once feeding a wild stag blueberries," he was appalled, his urge to protect her instantly activated. A stag is, specifically, a male red deer, the native deer of Great Britain, where the couple was living when Plath published "The Fifty-Ninth Bear." However, a question hovers behind the page, a question important to our discussion here: Is he right? A full-grown male red deer is an impressive, even fearsome animal, with a wide, spreading rack and, when confronted with another male, a guttural roar. A big one weighs over five hundred pounds. But is Sadie really in danger?

When I first found the story, I dismissed the supposed peril. It was Plath's opinion coming through, I thought, an opinion I do not disagree with: that by living in fear, cringing at the thought of standing within striking distance of a harmless deer, Norton impoverishes himself . . . something like that.

But then I start working on this book.

Switching back to Yellowstone—I always knew some of the deer here were dangerous. First, the word "deer" has a broader meaning than you may think. Deer are members of the family that scientists call the Cervidae. They are hooved animals that graze for a living, and the males of almost every species grow and shed antlers. That much is familiar, but now think more broadly. We have a number of animals that fit that description in Yellowstone. Specifically, we have four members of the family in Yellowstone and Grand Teton National Parks—white-tailed deer, mule deer, elk, and moose—and some of them get to be a good deal larger even than the wild stag Sadie was feeding. And more than a few of them really can hurt you.

Bull elk lose their antlers every year. Collections like this are made by humans who want to steal them.

Indeed, they are a headache for the government, and at times a serious danger for the traveling public. At places like Mammoth, where the local elk herd is a near-constant presence, it can sometimes be the major focus of the local rangers, who cannot be blamed for getting tired of their company, lovely though they are. The moose are fewer in number, but are the largest by far among this family of animals, and can kill with ease.

Some are smaller, though, and not the same kind of threat. In Yellowstone National Park, there are two species of deer that most people will recognize as "deer" when they see them, mule deer and white-tailed deer. You will note, on your map and as you drive, a number of place-names with the words "Blacktail" and "Blacktail Deer" in them; local people tend to treat "blacktail" deer and "mule" deer as the same creature, although elsewhere, more fine distinctions are possible. The whole subject is confusing, because blacktail deer and white-tailed deer both have white backsides; worse, "mule deer" are called that because they have big,

floppy ears like a mule—but so do all deer. For our purposes, we can drop the whole matter, because there are so few white-tailed deer in Yellowstone National Park that the rangers do not monitor them, and for a population number, just call them "scarce" and leave it at that. If you see a deer in Yellowstone, it is almost certainly a mule deer.

There are thought to be between 1,850 and 1,900 mule deer in the park, during the summer. Those on the northern range head downhill and across the boundary during the winter, following,

Antlers, if you look closely, are often gnawed on the ends. Rodents do that, for the nutrients.

A Gardiner mule deer. They will very nearly come in the front door.

This one is Big TJ, who is called big because she is pregnant.

roughly, the Yellowstone River to lower elevations and ultimately into Paradise Valley in Montana and the surrounding national forest. They can turn up anywhere during the summer, but the place to look is inside the town of Gardiner, where they can be found on either side of the river, eating people's lawns and shrubbery. They breed in November and December and have their fawns between late May and early August; this extended birthing season means that does will regularly be accompanied, on those Gardiner lawns, by fawns that are lovable to a ridiculous degree. Around places like Gardiner and Mammoth Hot Springs, and throughout Greater Yellowstone, you will see fences made of various kinds of galvanized steel mesh. They encircle shrubs, flower beds, and vegetable gardens and reach far into the empty sky overhead, as much as seven or eight feet. You guessed right: Those are there to stop the deer and elk from demolishing the contents. Even though they can be a great nuisance, the deer are a part of the town, and are regularly given names. Yes, they can be told apart; they often have a distinguishing feature, like the unusually prominent unibrow worn by the male who hung out in our neighbor's yard last year. They named him Rupert, but he looked rather like John Belushi.

And there, in that familiarity, lies the problem. Mule deer are surprisingly bulky, males weighing 150 to 250 pounds, and females between 100 and 175, even though they only stand three and a half feet at the shoulder. They are deceptive: They are adorable, yes, but they can kick and prod with their antlers, even though it isn't likely. That is one reason it is illegal to feed them in the park, and not a good idea outside, although people routinely do it. The deer get hooked on it, and bad things can happen—as with other animals, you don't want them loitering in a place where there are motor vehicles buzzing around. Furthermore, we tend to focus on the danger wildlife poses to humans, but our pets are much smaller than us, and they can get themselves into very hot water with deer. Some dogs react to elk and deer with indifference. Some want to tear them apart, absolutely freaking out at the sight, and little ten- or twenty-pound city-bred dogs taking on a six-hundred-pound elk do not know what they are getting into.

A pair of Gardiner mule deer fawns.

Their adorability itself gets them in trouble.

There is another danger here, however, one that is a good deal worse than any violence even a seriously annoyed deer is going to inflict, a microscopic hazard inside the bodies of these animals that is the most dangerous element in Greater Yellowstone, excluding only the supervolcano. It can be in the deer, and in the elk and moose, too. It is not a disease organism. It is not even alive. But it can kill in extraordinary ways, even though the animal itself has no control over this hazard, and the damage is done after the animal dies.

We will give it attention at the end of this chapter. In the meantime, let us look at how these creatures can hurt you when they mean to do it.

Oh, and the chapter title. It requires a little background.

The classic films that the Disney animation studios in Burbank made while Walt Disney was still alive were demanding projects to work on. Take *Bambi*, for instance, released in 1942 (the Bambi family, incidentally, are white-tailed deer). The animation work was famously grueling. Walt wanted realistic motion in the animals, and every twitch and jerk and blink they made had to be individually drawn, in thousands and thousands of separate cells. If you have seen the film, think of Bambi and Thumper on the frozen pond; all that had to be drawn *by hand*. The animators naturally wanted to squeeze in a little extra downtime, and would goof off at work as much as any office employee does, but then Walt would come walking down the hall in their direction, and they would scramble back to their stools and drawing boards. It was hectic. They therefore developed a code. When one of them spotted Walt Disney—the man, not the corporation—walking in their direction, he would call out, softly, "Man is in the forest!" The man was Walt, but the reference was to *Bambi*, in which mankind is the villain.

Think of the rest of this chapter as Bambi's revenge. Man is in the forest, all right, but the deer, in the following pages, will be taking a few trophies of their own.

When you start digging into a subject like this—the strengths and capabilities of all the forms of life that live here in Greater Yellowstone—you move from one wonder to another. The creatures are a mirror to the astounding landscape itself, as filled with marvels as it is.

When you meet one up close, you come instantly to understand that the moose very much belongs in the category of "marvel."

But you do not want to meet one up close, and this time it is not a matter of being excessively cautious in the face of a relatively remote danger, as with a buck mule deer. Moose really are dangerous, and up close, they can be terrifying.

The moose is the biggest member of the deer family, in every way: the heaviest, largest, and tallest. In North America, it is the tallest land animal, and in Yellowstone, we might have more of a sense of how unusually massive they are if they did not have to share the landscape with all those bison, which are the biggest mammal in North America by far. Like the wolverine, the moose is a circumpolar animal (although their range is somewhat to the south of the wolverine's). In the northern forests of the world, what ecologists call the boreal forest, through Alaska, Canada, the northern United States, northern Europe, Scandinavia, and all the vast sweep of Russia and Siberia, the moose finds its home. It spends its days there eating and eating, chewing and chewing, hour after hour, consuming twenty-three thousand calories in plant matter every day to maintain its terrific size. One old Yellowstone ranger had a theory that an important adaptation that an animal like that must possess is an ability to tolerate boredom.

From a distance—the only way to see them—that size is hard to get a fix on, but it is extraordinary. Let us talk about superlatives first, if only for the fun of it, and so turn to the subspecies called the Alaska or Yukon moose, *Alces alces gigas*. They are also sometimes called "giant moose," which is a literal translation of the Latin name, and which also simply fits. Males average fourteen hundred pounds, females a perfectly respectable thousand—that is, half a ton. Hunters, always, are focused on the antlers. Alaska moose antlers on average are almost six feet across. Imagine carrying that around on the top of your head; it is actually kind of horrifying. Given a choice, I suspect a bull moose would want his antlers anyway. He uses them to dominate smaller males, to spar with rivals, and for one other purpose not immediately apparent: The distinctive "palmate" shape of the antlers—shaped like an open hand—functions like the parabolic microphones that broadcasters use at football games. They gather sound and reflect it down into the moose's ears, improving his hearing significantly. So, they are worth it, even given what they weigh: in Alaska, between forty-five and fifty pounds.

It is just as well that moose do not know what we think about them, because it is difficult to look at one and fail to note how gawky they look, how lumbering, ungainly, uncouth even. There is a good reason: It is an effect of us seeing them regularly when they are not in their preferred element. They are a little like those dinosaur skeletons—the really big ones—that were hard for the early scientists to understand because they had to learn to conceive of this giant body suspended in water. That is the moose's preferred environment; they are happiest in water, which is what their body is best suited to. That distinctive, outsized nose is perfect for eating submerged vegetation, because it has nostrils that can be closed to shut out the water. The hooves are adapted for water, too, or for muddy bottoms; the hoof splays when the moose steps down, creating a snowshoe effect that works well on the soft bottoms of marshes and lakes. The outer layer of the moose's coat consists of guard hairs that are hollow and so help the moose float while swimming (Thidwick the big-hearted moose, you surely recall, escapes by dumping his guests and swimming away to the far shore of Lake Winna-Bango). Its long legs are also ideal for a life spent in the water—and the snow, which has to be deep indeed to defeat the moose. Some other features of that body, though, like the dewlap under the jaw, remain somewhat mysterious. In the case of that dewlap, swinging like the clapper in a bell, the effect is to increase the aura of moose-ness considerably. Moose are formidable, but will never be mistaken for elegant. In the deer family, they are the country cousins.

They are certainly not to be trifled with. The Yellowstone subspecies of moose is, it is true, smaller—but not by that much. An adult bull weighs not quite one thousand pounds, and a cow perhaps a hundred pounds less. At the shoulder, they stand between five and a half and seven and a half feet tall. The subspecies, sometimes called the Yellowstone moose, bears the formal name *Alces alces shirasi*. It and Yellowstone have a history together. It is named for George Shiras III, a Pennsylvania attorney elected to Congress in 1902 but today remembered best as a pioneer of wildlife photography, especially in the use of flash photography to catch quick-moving animals during an era of slow-moving camera shutters. Shiras (usually pronounced *Shy*-rass) worked mainly at night, performing important early experiments with the camera trap, a method of waylaying his subject by anchoring a camera beside a path frequented by the animal that he was after. The shutter was activated when the animal tripped a wire and lit off a pan full of magnesium, which must have been quite an eye-opener. *National Geographic*, at that time a sleepy academic journal, surprised everyone with their July 1906 issue, in which it published seventy-four of Shiras's photographs, in one step going a long way toward turning itself from a minor publication into a media powerhouse. The magazine, in a 2015 tribute to Shiras, called him "Grandfather Flash."

In 1908 and 1909, he came to Yellowstone National Park looking for moose, among other things. Shiras had heard that the Yellowstone moose looked different; it was smaller, lighter in color, and had a face that did not look right to one accustomed to seeing moose elsewhere. He went out and found them. The park moose was in fact a new subspecies, which was later named in Shiras's honor, *Alces alces shirasi*, the name it bears to this day.

In Yellowstone, the moose has generally not been a threat to humans, mainly for the simple reason that there are not that many of them, perhaps two hundred in Yellowstone National Park itself, although it is hard to count them because of their habit of sticking to deep cover and keeping to themselves. They do not form the kinds of large groups characteristic of, say, the elk you will see everywhere in squad- and platoon-size clusters. In the early years of the park, there appear to have been no moose at all. The first documented report of one on the northern range occurred in 1913, when Yellowstone National Park had existed already for over forty years; when Shiras went looking for moose, he did so to the south, although they were scarce in Jackson Hole, too. The population expanded because hunting was controlled, predators were killed off, and forest fires were suppressed; Yellowstone moose prefer forests of mature spruce and fir trees, and so fire suppression and Smokey the Bear helped them—and then the massive fires of 1988 led to what almost amounted to a crash. Moose numbers went from about a thousand in the 1970s to the low of two hundred that continues today. There is some evidence that this is more or less normal: Their numbers grow, and then they go in the opposite direction.

Moose can certainly be aggressive. They have an ability many hooved animals do not possess, a kind of ungulate superpower: They can kick in all directions, not just behind. Moose are especially a problem for people in Alaska, where they actually injure more humans than do bears. The Alaska Department of Fish and Game, in their newsletter, even ran a whole article about the moose's talent for kicking; it is a common enough occurrence there to warrant such coverage. "Unlike bears, moose don't bite," the article explains. "They kick. In a moose's world, where bears and wolves want to eat you and are especially eager to eat your babies, kicking and stomping can be a formidable defense. Sometimes people wind up on the wrong end of those hooves, usually belonging to a defensive mother moose."

The article was written by Riley Woodford; he spoke to a wildlife biologist from the Kenai Moose Research Center, John Crouse, who understands the kinesiology and physics of the moose kick in some detail. "They generally kick with their front legs," Crouse said. "That's most common, but it depends. Sometimes they get annoyed and they'll kick back. If they feel threatened, they can kill you with a hind leg kick." Their hooves are sharp and pointed, too, so they are kicking with feet that are deadly weapons in themselves.

Woodford continued:

When a moose delivers a hard, fast kick with its hind leg, Crouse said you can hear the tendons and ligaments pop and snap when the hoof comes back at the end of the kick. They can kick straight ahead and will also rear up and stomp.

"The front leg will come straight up with that hoof, they'll also come forward and kick out at the side, and they can come down as well, on to the top of your head or shoulder," he said. "From what I've observed they can pretty much kick 360 degrees around them."

It's common for moose to kick each other. There is a hierarchy in a group, and that order might be stable for a period, but Crouse said it's dynamic. It changes with the season, their mood, and whether a cow has a calf. Outside the breeding period conflicts are mostly associated with food, he said.

"They're browsing and one moose wants a particular shrub. Often, they have it sorted out, you'll see their ears lowered and other forms of communication, but if they don't have it worked out, they'll be kicking, or standing up on their hind legs and boxing each other. They'll do that occasionally to us, rear up on their hind legs. Instead of being six or seven feet tall at the shoulders, they're nine feet in the air, and their legs are above your shoulder. Their head is well above you; and the hooves are above your head."

Moose get frustrated and Crouse described what he called moose tantrums.

"They'll get so upset about something sometimes they have to blow off that energy," he said. "They'll run around in circles, their hair is all erected and up straight; they will run with their legs jerking out and kicking sideways."

It is a pity we have so few moose in Yellowstone. That would be quite a sight on, say, the big lawns at Mammoth Hot Springs.

Moose do attack people with some regularity, in places where there are enough moose around that people can run afoul of them. In March 2024, moose attacks were in the news after an incident during the Iditarod sled dog race in Alaska. The racer who went on to win had to first kill a moose that attacked and injured one of his dogs. Similar incidents involving Iditarod sled dogs and moose have happened repeatedly, and the race has rules that specifically cover such situations (the racer who killed the moose in 2024 was penalized for improperly gutting the animal—yes, they have rules about that—but won anyway). After the most recent Iditarod incident, the *New York Times* reviewed the evidence on moose attacks around the world, and reported that a final accounting, at present, is not possible:

> *Moose attacks against humans are uncommon, but they do happen in states with large moose populations. In Alaska, for example, as many as 10 are reported each year. And in Colorado,* The Colorado Sun *reported in October that at least four people were injured in moose encounters in 2023.*
>
> *Ironclad statistics about fatal moose attacks against humans are unavailable, but wildlife experts agree they are extraordinarily rare as human fatalities involving moose result almost exclusively from vehicle crashes. (One case of a fatal moose attack in Sweden was originally thought to have been a homicide, according to a 2017* Journal of Forensic Sciences *report.)*

Working for a big university, I have access to things that are not on the wider internet, including that report in *Journal of Forensic Sciences*. It permits me to declare, here, that when they want to be, moose are every bit as dangerous and catastrophically destructive as grizzly bears—or great white sharks, for that matter.

The incident happened when a woman took her dog out for a walk in a rural part of Sweden. When her body was found later, investigators, as the *Times* noted, thought they were dealing with a murder, and it is revealing to learn what they thought the murder weapon was: The police started with the theory that the case was "homicide by means of a riding lawn mower." Yes, they thought she had been run over with a riding lawn mower.

It was that bad. She had died when extreme force essentially caved in her chest, a condition doctors call flail chest, but one suspects she would have died even without it. She was cut and contused literally all over, and bones were broken all over, too. One leg was more or less completely mutilated. There was a distinctive and massive bruise across her torso that the investigators said was "consistent with a heavy stomping or crushing force" applied when she was pressed against a log on the ground. Most shocking, oddly, was not the injuries, but the condition of the pants she was wearing. They were denim jeans, and to judge from the photos (the damage and bloodstains make this partly a guess), they were Levi's 501s. They were, that is, denim pants known for their sturdiness. This pair had been torn like paper, from the waistband down the right leg well past the crotch, the leg that had been shredded by lacerations.

So that is what moose can do with those sharp hooves and antlers. The investigators had collected hairs from the body, and after they abandoned the amazing lawnmower hypothesis, they had a close look and discovered them to be moose hairs. DNA evidence, once they knew what to look for, was all over the body—moose DNA, and despite the lack of witnesses, the investigators were now able to put together a theorized sequence of events, their version of what probably happened. Recall that she was out walking her dog. One major reason the rangers won't allow you to take your dog into the backcountry in Yellowstone and Grand Teton National Parks is that the dog will see an animal and either want to play, or fight it to the death. The animal—this can happen with bears especially—will whack the dog like a badminton birdie, and the now-terrified dog will run, leading the now-furious animal straight back to you.

Something like that happened here, too: "The woman was walking the unleashed, young dog when the dog came across a moose and returned to the woman, seeking her protection. Probably, the moose pursued the dog, knocked the woman over, and killed her by stomping, kicking, and goring." The rip in the pants happened when the moose's antlers by chance slipped under the waistband of the Levi's. The waistband then became a handle, and the moose used the leverage to lift and throw the woman through the air.

The moose must have been in an extraordinary state of mind, a maniacal rage that is, of course, not their normal state. Picture a moose in your mind, and you may see Bullwinkle first, but then, perhaps, the actual animal, standing still in a meadow, placidly chewing and chewing like a hillbilly in an old cartoon, chewing tobacco. You half expect them to spit at a spittoon. You got it right; left alone, they are harmless. That, combined with the small population of moose, means that moose attacks have been vanishingly rare in Greater Yellowstone, even with an annual visitation, for Yellowstone National Park, that is now around five million people.

It certainly does happen. It happened a few months ago, as I write this, although the incident did not have a particularly serious outcome. In May 2024, a video was posted to Instagram that showed a group of people walking off the road and into the meadow at Round Prairie, a lovely location in the Northeast Entrance Road corridor, near the Pebble Creek Campground and the junction of Pebble Creek and Soda Butte Creek (the latter is one place where moose do turn up regularly, along with Pelican Creek, the Gallatin River, and the Lewis River). Contrary to what a lot of people seem to think, leaving the road is not actually illegal, but what some of these people were doing was: Apparently, they were trying to get a closeup of a moose. The video was taken from a great distance, and it is not clear what happened, but two moose did take off after the photographers. It was a thrilling sight, actually, the two moose running as if at the Kentucky Derby, the humans running for their lives. Nothing came of it, although a funny feature of the video is the voice of one of the men in the group that was taking the video, back on the road. He was cheering for, yes, the moose, urging "Get 'em! *Get 'em!*" A day spent touring the park and dealing with the traveling public has that effect on a person.

There have been other incidents. A much more serious one occurred back in 1990, when a woman named Beverly Buettner, who owned a health food store in West Yellowstone, was attacked not far from town. The Salt Lake City *Deseret News* reported that the attack "left her with three broken ribs and a punctured lung. She said the moose saw her as she was walking along the Madison River, charged, knocked her to the ground and kicked her several times in the head, back and chest before fleeing." A serious matter, but Buettner blamed herself: "'It was a place that belongs to the wild animals, and I was trying to get out of there as fast as I could as soon as I realized there may be an animal nearby.'" She just did not move fast enough. Attacks actually seem to be a little more common around Grand Teton National Park. In 2023, Jackson Hole Radio reported on the most recent attacks there:

> *According to Wyoming Game and Fish Department officials, the moose population in Wyoming is just under 3,500 animals, with an estimated 276 living in Jackson Hole according to Wyoming Wildlife Advocates. These moose can be found in the Willow Flats, Moose Wilson Road and along river corridors in Grand Teton National Park during the summer, and by the sagebrush flats during winter.*
>
> *Moose attacks are relatively rare in Jackson Hole, and tend to happen at least annually. This trend has been followed since 2021, as there has been one notable incident a year.*
>
> *During 2021, a moose forcefully knocked a man to the ground while he was on the bike path northeast of Wilson Elementary School, resulting in his hospitalization. In 2022, an 89-year-old man named Ed Opler was knocked to the ground by a moose, causing him to suffer a broken scapula. In March of this year, 27-year-old Hannah Garland was charged, head-butted, and stomped on by a moose. The attack led to severe injuries, including a concussion, six broken ribs, and bone bruising on her right arm and elbow.*

There have been more, of course, but the records are spotty. Locally, they are not a major issue.

Still, you may find yourself face-to-face with one of these awesome creatures—and actually, you will not be face-to-face, because the moose will probably be taller. It happened to a friend in the early 1990s: He turned a corner in dense forest, in a part of the park where moose do not turn up much, and there he was, not six feet away. My friend went back around that corner in some haste, and that is what he was supposed to do. If you have a run-in yourself, watch for the moose's body language, some of which you might be able to interpret instinctively. If the animal's ears are pinned to its head, it is annoyed. You do not want the moose to be annoyed. If the hair on its neck and shoulders is raised, if its eyeballs are bulging, and if its tongue is hanging out, the animal is angry with you and

A moose with her calf. Mom can be dangerous indeed. GETTY IMAGES

may attack. A cow moose with a calf will charge to protect her offspring, same as a bear. If one does come after you, try to get behind a tree or some other kind of obstacle. Do not leave the moose feeling as if you are blocking its avenues of escape. It wants to get away from you, or for you to get away from it. If you fall, curl up into a fetal position; protect your head with your hands and arms, and stay that way until the moose leaves.

Problems mostly arise from people being silly. There is one other Greater Yellowstone moose attack to report, although it is not clear if anyone was injured. It was recorded in a video posted on Instagram on April Fools' Day 2023. In the video, the potential victim, scrambling for his life, and the pursuing moose disappear behind a parked SUV, and no one seems to know what the final outcome was. It may have been painful, but cannot have been serious, or we would have heard more. The incident took place outside a bar in Big Sky, Montana, the ski resort community just northwest of Yellowstone National Park. In the video, a pair of thoroughly inebriated young men approach a moose, which has its back to them and is partway up a snow-covered slope. The man taking the video, Jake Hopfensperger, is trying to get them to desist; my guess is that he was the designated driver that night, and designated walker, too. In defiance, one of the drunks taps the moose on the rump. The animal at this point has had enough. Ears back and head down, the very picture of cervid rage, the moose turns and charges. The drunk turns himself, but the parking lot is covered with ice. He slips, tries to rise, slips again, and can only scramble on the impossible footing like a man in a nightmare, while the moose, which has no such trouble, runs him down. That is when the two disappear behind the vehicle.

But there is one more bit of dialogue at the end. Hopfensperger, who has had enough himself, shouts the words that seem to come naturally, the universal language of those who have been pressed beyond endurance by silly people bothering the wildlife:

"Get him!" he says. "*Get him!*"

We are now down, in this chapter and the next, to our last two large animals—but they are large indeed. Pound for pound, they may cause more trouble for humans than any other, and represent what is arguably the greatest threat to life and limb. They are also among the most magnificent creatures we have in our already-gifted lineup.

Yellowstone National Park has long had a single iconic animal that captured the spirit of the place. For decades, it was the bear; it may now be the wolf. However, a visitor who spends some time here may come to object to that choice, especially if the visit lasts longer than the single day or two that most people have. Give it a week, and other creatures demand attention and respect, especially if you stay at Mammoth Hot Springs, and especially if you come here in the fall. Then, the star of the spectacle is the elk.

The moose may be the largest member of the deer family, but the elk is not far behind, especially in Yellowstone, where the moose are not as large as elsewhere—and here is a source of terrible confusion to visitors from Europe, where the word "elk" refers to what we call moose (North American officialdom has been trying to get people to call them "wapiti" for decades, a Cree and Shawnee word meaning "white rump," with little success). The elk, *Cervus canadensis*, is the second largest of the deer family, but bulls can weigh seven hundred pounds in Yellowstone, about three hundred pounds less than a Shiras moose. As with moose, male elk are bulls, females are cows, and the little ones are calves; young males are called "spikes" for the ten- to twenty-inch unbranched and sadly adolescent-looking antlers they grow. Cows are about five hundred pounds. The bulls are five feet high at the shoulder, the cows a little less. In a place where the two species mingle, like Gardiner, you can get a sense for how much larger the elk are than the mule deer. They are actually easy to tell apart, once you know one trick: Elk have dark brown fur on their necks and heads, light brown on their bodies, and the line between the two is exact.

As noted, Mammoth is a good place for elk. Elsewhere, they tend to disperse upward into areas where the grass is good later in the year. They come back down in the autumn, as part of a long movement away from the coming snow, and so move back toward the roads and the villages—which means that they make themselves scarce in developed places during the height of the visitor season, and reappear at the exact time the human visitors are going away, an irony that the employees do get

The Roosevelt Arch with a cow elk, a member of one of the Gardiner elk gangs.

a kick out of (give them a break—the height of visitor season is a great trial). The open lawns at Mammoth make it easy to follow every move; Gardiner is similar this way, because it has the school athletic field, which the elk love. At Mammoth, they are always around, and always have been. The soldiers here, when Mammoth was a military outpost, were watching them around the turn of the twentieth century, same as we are today, and I have always suspected that our elk are direct descendants of theirs. Again, look for the big "deer" colored in two tones, with a distinct boundary on the neck. The large elk, the ones without antlers, are (mostly) the cows; the small ones that follow them around are their calves. Depending on the time of year, you may see a calf nurse. The bulls are generally not around—until the autumn, when things get exciting. That is when they enter the rutting season, or simply "the rut."

Two-year-old bulls have thin antlers with four or five points on each side; when they are three years old, they still have four or five points, but the antlers grow thicker. At that age, their antlers have nearly all the qualities they need to accomplish the purpose for which they are intended.

A Gardiner cow elk.

Four-year-old bulls will have the famous "six by six" or "six-point" racks, with six separate termini on each side of the antlers. They will now, if lucky, grow six-point racks every year until they are eleven or twelve, when they achieve the heaviest antlers of their lives. After that age, their antlers will be less and less impressive . . . if they survive, which becomes more and more unlikely. Every year, their bodies put an amazing effort into this display, only to cast the antlers free in March or April and start over again in the spring.

In the autumn—and this is one of the spectacles you can see best at Mammoth—the mature cows go into estrus. The bulls know this, and begin to bugle at each other. There will, in a given area, be one bull that is at the top of his form, with the largest and heaviest set of antlers and the largest and most powerful body. The cows—they are free agents in this, except that they are enslaved by their hormones—move toward his bugle. So do other bulls—the bull has different bugles, some for cows, some for other bulls. The cows will form themselves into a harem that is under the rough control of the dominant bull. Other bulls will approach, but as is usually the case in nature, the bulls do not fight at the drop of a hat. They look each other over, and that one bull who is most fit is so plainly the dominant animal that the smaller bulls will back down, will indeed flee, if not far. They will then become "satellite" bulls, hovering around the cows like moons around a planet, hence the name. It is a wretched sight, because the satellites are so desperate, but the dominant bull will not let them get close to the cows. The cows do not necessarily object, and while the big bull is chasing a rival, one of the younger bulls may rush in and take advantage of his distraction. Mostly, though, the satellites have to wait for that future year when they are the ones bugling in the cows and dominating the youngsters.

The bugle is an extraordinary sound. One expects a call like a trumpet might make (after all, Mammoth is also Fort Yellowstone, a former cavalry fort). Instead, the bull makes an ear-splitting, quavering, falsetto shriek that carries a great distance. It draws in all possible rivals, and sometimes it draws in a bull that is the approximate equal of the animal that issued the challenge. There will then be a fight, a protracted wrestling match with antlers locked together and knocking on each other with a

A cow and bull elk, brought together by the rut. GETTY IMAGES

sound like pieces of wooden dowel dropped on a hard floor. The cows and last year's calves look on, awaiting a winner. The fights sometimes go on and on, and can be literally exhausting. In the end, there will be a winner, who will bugle like the tyrannosaur in *Jurassic Park*. Bulls, however, lose up to 20 percent of their body mass during the rut, and finish it covered in injuries. The strain can be fatal. The good news is that the cows will, by October, be pregnant. The bulls return to their usual life as solitary wanderers, and the cows re-form their small herds, like the one that has always moved around and among the buildings at Mammoth Hot Springs. A friend used to call them "The Ladies' Cotillion."

The Ladies' Cotillion.

That is what you can look forward to if you can spend time at Mammoth in the autumn, generally in the second half of September and into October. In a book he published in 1984, *Mountain Time*, the former ranger and historian Paul Schullery commented on life at Mammoth, where he lived and worked. He spent many hours watching the elk, and had occasion, sometimes, to feel an emotion less like delight and more like worry. The people who live there have a different perspective on the elk, which they have to watch not for fun, but as a way of keeping out of the hospital:

> *For most of the year, except in the heat of summer when they move up country to the flats south of Bunsen [Peak], elk are a daily event at Mammoth. They graze on the lawns, bed down between the houses, and generally hang around—not tame, not too familiar, just there, part of the setting. It is hard to explain how routine a part of daily life they become. For most of the year you just have them around, like so many 600-pound robins on the lawn. Seeing them daily, learning to recognize individual animals, occasionally shooing them (never without a twinge of fear) out of the way like uncorralled cattle—all this contact makes them common in the mind; appalling proof that civilized people can become blasé about absolutely anything, even huge wild elk munching the flower bed.*

Park Service regulations require park visitors to keep a distance from animals. The general rule prohibits "Willfully remaining near or approaching wildlife, including nesting birds, within any distance that

"Man Is in the Forest!"

Downtown Gardiner. Yellowstone has helped create a new kind of animal—the urban elk.

disturbs or displaces the animal." That applies to every species. More specifically, "Park regulations require visitors to stay a minimum of 25 yards—the length of two regular school buses—away from most large animals and a minimum of 100 yards—the length of a football field—away from bears and wolves at all times." This is the one that applies to elk: no closer than twenty-five yards, and it makes sense.

But Schullery actually lived at Mammoth, and when you live there, you end up bumping into the elk when you do not want to, not at all:

Usually it is easy enough for us—people and elk—to overlook the other's presence. The elk keep a distance that varies with their mood, and we pass each other with only mild wariness. If several are bedded down along one of the paths in the woods around Mammoth, they'll usually stay there as I pass. Like domestic stock, they're inclined to scatter if rushed, but nobody sensible about the terms of cohabitation stirs them up. Most of us have encountered less tractable elk at one time or another, one that stands her (it seems to be a cow most of the time) ground mid-path as we approach. More than once I've been walking up the path to my door and, glancing casually over at the nearest elk, have seen a restlessness in her demeanor that signified an elk in a bad mood. She must have seen alarm in my glance; she glared at me and something set her off in a quiet rush in my direction. I didn't wait to see if it was a bluff. As an early park visitor said when describing how he'd escaped from a bear, "I jest natchully faded away. I reckon that atmosphere is all het up yet with the way I come through it."

In this matter, the language is not doing us a favor. The words "bull" and "cow" lead us astray; we assume it is the bull that is the dangerous member of the nuptial pair, since in the barnyard, the bull is the one who will kill you, while the cow allows herself to be placidly led to the milking machinery. But Schullery is correct: It is the cow elk that are the problem.

In 2022 and 2023, we had a cow elk in my part of Gardiner that was an outright terror. My neighbors, the ones who named Rupert the mule deer, never came up with a name for this one, calling her, instead, simply "the crazy one." She had to be watched, although watched with care, because you did not want to make eye contact. If you did, she would

Bruce, the family dog, greeting a visitor at the front door—a cow elk.

come at you. She pinned my mom against a fence. She pinned an old friend from the days when I worked in the park against a fence, the same cow, the same fence. After escaping, the friend called me, breathless with fear, while the mad cow held her husband literally prisoner in his vehicle. He told us later that he was not sure what to do about it; his keys were in the house, and he was in the vehicle, with the elk staring him down (apparently, she could see through the glass). When he tells the story, he always notes that there was a long, ropey strand of drool running out of her mouth. We should have named her; my vote was for "Lizzie Borden."

The cows are, of course, protective of their calves, and the bulls can be perfect terrors, too. Mammoth is again ground zero for elk trouble as much as for elk thrills. During the rut especially, the rangers have a long, exhausting ordeal on their hands. You will see them, wearing reflective vests, usually more than one of them, holding the humans back from the magnetic pull of the elk. During the rut, when the dominant bull is patrolling the boundaries of the harem, the satellite bulls are desperately, *desperately* looking for an opening, and the cows, with last year's calves, are inclined to wander off, placing even greater demands on the already-frantic bull. It can get to be more of a temptation than the park visitors, cell phones at the ready, can resist. The G-men and G-women have recourse to one trick when things begin to get out of hand. The trick: Take a spade-type shovel and drag it across an old asphalt road or parking lot, so that the blade rattles on the bits of gravel that are exposed on the surface. It makes a ringing, tinkling sound that is more or less neutral to humans, but is the elk equivalent of fingernails on a chalkboard. When the rangers need to get the elk to disperse, that will do it; the elk freak out and move off quickly, with a maniacal look about them. Let the rangers do the spadework, though; I am revealing the secret here only because park visitors rarely have twenty-inch D-handle spade shovels in their luggage.

Bruce is unusually even-tempered. Most dogs would not be so sedate around a visitor such as this.

Given the situation, there is bound to be trouble, and every year there is plenty. When the elk get aggressive, it is not just other elk that are hurt, and not just people, either. When the rut began in 2014, the NPS issued a press release: "The fall season bull elk rutting activity has begun. Bulls are much more aggressive towards both people and vehicles this time of year and can be a threat to individuals and property. Several vehicles are damaged by elk every year and occasionally people are charged by elk and can be injured." It is not at all clear what the elk make of vehicles; there is a tendency to believe grazing animals are naturally stupid, but they surely know the difference between a rival bull elk and a Ford Explorer. Whatever they are thinking, bull elk charge and gore vehicles with some regularity. The world of elk and human intermingle, endangering both sides in the most unexpected ways:

> *Area residents are reminded that during this period, it is not uncommon for bull elk to mock fight with many types of household items found in residents' yards. As a consequence, bull elk often get household items wrapped around their antlers. This can result in bull elk getting tied to each other, or to brush, trees, or other objects which can ultimately lead to their death. Over the last few years, bull elk have had to be captured to remove extension cords, cloths lines, shrubbery baskets, leashes, wire, nets, cloth bags, swings, hammocks, coaxial cable, and badminton nets (complete with poles) from their antlers. During the fall rut, residents are asked to make an effort to remove all such items from their yards when not in use.*

That sounds comical, and it does have a funny side—but not from the point of view of the elk. Schullery describes a Mammoth elk that got his antlers tangled up in Christmas lights and wandered the area "as if he was looking for an electrical outlet." He was tranquilized and the lights removed—but while he was still tranqed and helpless, he was threatened by a coyote pack.

He survived, and may have had his revenge. Elk threaten and charge people with great regularity. Much more rare is actual injury (we will leave out injuries caused by automobiles hitting elk). In 2018, there were two incidents in three days, both, unsurprisingly, at Mammoth. On June 3, Charlene Triplett, a fifty-one-year-old from Las Vegas who worked for the park concessionaire Xanterra, was walking behind the

Bull elk, after the rigors of the rut, often do not make it through the winter.

Mammoth Hot Springs hotel when a cow elk came at her. The cow is thought to have been protecting a calf that was bedded down twenty feet away, behind a row of cars and out of sight. It was a serious matter. The elk did what cow elk do: She reared up and back like a horse in a Western movie, an astounding sight—especially, I suspect, when you are underneath it. The elk then kicked with its front legs, hitting Triplett in the back, torso, and head. Her injuries were serious enough that she was flown by air ambulance to the Eastern Idaho Regional Medical Center in Idaho Falls, a Level II Trauma Center with a Level I ICU, about the best care she could have gotten in this surprisingly primitive part of the world.

It happened again on June 5, and again at Mammoth; no one knows if the same elk was involved, but given that it was almost the same place, and given especially that elk have personalities—in this case, a violently protective personality—it seems a fair bet. The victim was a woman from Cypress, Texas, and because of the Texas connection, the *Midland Reporter-Telegram* talked to her. It was the kind of situation in which no one is really to blame. Things like this are going to happen, when you have a five-hundred-pound mother elk with an attitude living side by side with the traveling public:

> *Penny Allyson Behr, 53, was walking from her cabin to the community bathroom at Mammoth Hot Springs Hotel when she turned the corner and found herself nose to nose with a cow elk.*
>
> *"We startled each other. I'm 5 feet 10 inches tall, so it was a big animal," Behr said.*
>
> *Behr said she didn't know what she was supposed to do, so she took two steps back and quietly greeted the animal. When the elk, which had been grazing with its calf along a cabin wall, took two steps forward, that's when Behr said she got scared.*
>
> *"I turned partially to look for a place to hide, but when I looked away the elk reared up on its back hooves and came down on my shoulder and collar bone. I remember screaming, and then I was on the ground," Behr said.*
>
> *She was then taken to the area hospital by ambulance, but was released the same day. Her injuries included a few broken ribs, a fractured clavicle and a concussion. She also had a few gashes in her head, but no internal damage.*

Behr continued her vacation, although the trip took a turn toward the unambitious. The rangers took no action against the elk, except to haze her and the little one away.

After that attack, a park spokesman told a reporter, "We don't have a system that tracks these incidents. Anecdotally, elk injuries are not common, although elk charging people seems to happen each year. Not all of those instances are formally reported to us." It actually happens quite often, maybe once a week, or maybe every day. And it is often difficult to define what constitutes an "attack." In 2018, an incident occurred at Mammoth in which a park visitor was knocked down by a bull. In the inevitable video, the elk moves quickly toward a group of people who,

"Man Is in the Forest!"

Elk mother and child. The calf is five days old.

it must be said, should have headed for cover instead of standing there holding up their phones (did Steven Jobs have this in mind when he oversaw development of the smartphone?). A ranger had been warning them away. Once the video was posted, everyone jumped down the victim's throat, ranting about how *stupid* he was . . . and then it developed that he was a foreign national, from Italy, and never understood what the ranger was saying.

There is almost a genre on YouTube and elsewhere that can be thought of as the "Raging Bull Elk" video. If you doubted me, above, when I said elk attack motor vehicles, these videos will show you what I meant. One that was recorded on September 10, 2021, is representative. There is, at Mammoth, a place where the roads make an oval, with the post office and clinic on one side, and the courthouse (the Justice Center is its formal name) on the other. Roads radiate off the oval, running toward the campground, the visitor center, the hotel, and Gardiner, and in the middle of all these exits, that oval makes a de facto traffic circle.

In the center is a bright green lawn that the elk like, and might like even more if they were not, essentially, on stage when they come here. The lawn will hold roughly an entire harem—there are fifteen to twenty cows in a typical harem, at its largest, although the number can run up to thirty—and there are times when the whole harem is on that lawn, the bull patrolling the edges, with cars accumulating on the asphalt until the elk are entirely surrounded by motor vehicles. Again, the best analogy is with a Western movie, in this case the traditional circle-the-wagons scene; here, though, the image is reversed, with the wagons circling the restless natives.

In the September 10, 2021, video, the bull is on that lawn alone; the harem begins to follow him toward the end of the recording. He is, I swear, about as worked up as I have ever seen a bull elk get. He digs at the ground with his antlers, chewing up pieces of turf. The antlers—just in and of themselves, independent of the seven-hundred-pound animal behind them—are formidable weapons, weighing thirty pounds or more. The bull paws the ground with his right front hoof, then his left, four times each, tearing up divots of turf and throwing them backward. And then he charges. Over and over, he does it; he reaches a point at which he finds the mass surrounding him intolerable, and he charges—but not at people. They are all well away from the bull, or are inside motor vehicles. That is what he is charging: motor vehicles. Over and over, he drops his head and makes a run toward his challenger on the road. How much, I wonder, did the heaviest of those vehicles weigh?

That is actually a common theme: Some bulls just cannot abide vehicles, and the chief, dominant bull at Mammoth regularly seems to be one of those bulls. In another video, a bull takes on a rental four-door with Utah plates—and he *wins*. The sedan rental car flees. In another, he goes so far as to attack a ranger car. Given that we are in a Western movie, it appears that the elk are the outlaws.

Every year, it begins anew, and it is only a matter of time before there is another serious injury. You cannot have a national park, not a real one, without letting the animals run around loose. Many species are comfortable living beside humans. Plenty of others live at Mammoth: ravens, magpies, mule deer, Uinta ground squirrels, bull snakes, even badgers, and

I could continue at some length. Honestly, the government is doing what it can with a difficult situation. You can help them out, and help yourself, by doing the obvious: Keep twenty-five yards—the length of two school buses—between yourself, the family, and the elk.

And to think—we have not even touched on the greatest danger they pose.

New Guinea is the second-largest island in the world, after Greenland, and it remains, even today, an extraordinarily primitive place. Were it not for the Second World War, it would be even more a mystery. During the war, it functioned as a natural barrier between Australia and Japan's new possessions to the north, and so was the setting of a prolonged struggle, with the armed forces of Australia and the United States on one side, mostly, and those of the Empire of Japan on the other. In the struggle, the native people of the island ended up playing a role, rendering crucial aid to the Allies especially. By the time the war moved to the north, the island had been penetrated, to some degree. It remained, and remains, a forbidding place, with a spine of mountains that regularly run over 13,000 feet—mountains comparable to the Tetons. Its highest point is Puncak Jaya at 16,024 feet, a mountain with glaciers on its flanks, in spite of its setting; down off the mountains, New Guinea is a land of impenetrable, grueling tropical rain forest, a place where only the locals can ever thrive.

And even they have a rough time. During the 1950s, when the eastern half of the island was governed by Australia, the colonial authorities made efforts to better penetrate the highlands and bring medical care to groups that had never known Western medicine at all. As they did, they found, among the Fore people (usually pronounced "foray") in the Eastern Highlands, a peculiar disease. That disease earned a Nobel Prize for the doctor who figured it out, D. Carleton Gajdusek, because sorting that one out led to much more. The disease was called "kuru" by the locals, for their word to describe tremors or shaking; people with the disease gradually lost control over their own bodies, and in the end died. Kuru is what today is known as a transmissible spongiform encephalopathy. (Knowing a term like that is a good way to frighten people at cocktail parties, so you should learn it.) It is "transmissible" because it is, as Gajdusek figured out, not genetic; it was something the Fore were getting from their environment. It is a spongiform encephalopathy, because it quite literally turns your brain into a sponge: "Encephalopathy" is a disease of the brain, and "spongiform" means what it looks like. You have probably heard of a closely related disease, even though you may think you haven't. It is bovine spongiform encephalopathy: mad cow disease, the deadly brain condition that spread through cow herds in the 1980s and 1990s, with the United Kingdom as the epicenter.

Kuru is similar, but has a peculiar history. It is an obscure disease afflicting a tribe of about twenty thousand people who could not live in a more remote and inaccessible place. Furthermore, while it killed many Fore, especially women and children, up to and through the 1970s, the death rate slowed dramatically afterward, and today, the disease is nearly extinct. Still, kuru is widely known today among people with a scientific education, and also among people without a trace of a scientific education, for two reasons. One, while the disease is obscure, the microscopic entity behind kuru is anything but, and that fatal entity could potentially threaten anyone. We will get to that shortly. Two, it has a cause so gruesome that the details are actually intolerable. I have the details open on another screen here, in Shirley Lindenbaum's 1979 book *Kuru Sorcery: Disease and Danger in the New Guinea Highlands*. Lindenbaum, an anthropologist who worked with the Fore, is probably the leading expert in this matter. The details are, however, so *very* gruesome that I am going to make an exception from my usual practice, and censor myself. The key detail is that the Fore practiced what anthropologists call endocannibalism, in this case mortuary cannibalism. When a member of the family died, the rest of the family ate the corpse. The men ate the muscles, and the women and children ate the brain. That fatal entity is concentrated in the brain, so the women and children were the ones to develop kuru and die, after an incubation period that lasts as much as fifty years—that is, the victims were exposed, and it took fifty years for the entity to do its work. The colonial authorities, aghast, suppressed the practice of cannibalism way back in the middle of the twentieth century, but elderly people still die of kuru here and there, kuru that they caught from the brains they ate decades ago, because it takes so very long to manifest.

And what is this entity? That is a crucial question for us—because we have it here in Yellowstone. No one knows how widespread it is, but it is carried by the elk and other species of deer (but not, happily, by bison). It is Bambi's revenge carried to the ultimate extreme.

It is not a bacterium, or virus, or parasite. It is, in fact, not alive at all. It is a protein, the basic building block of life. This building block is not shaped right, and that destroys the cell of which it is a part. It is called a prion (usually pronounced like the Toyota car: *pree*-on).

For most people, "protein" is steak and eggs. It is what you are supposed to eat when you are on the Atkins diet, or (rather more obviously) the carnivore diet. For a scientist, however, protein is much more. Proteins are molecules of great size and complexity that do the work a cell needs done. Proteins also regulate the organs and tissues of the human body (and other bodies), and are required for their function and structure. They are made up of amino acids that are connected to each other in long chains. Their sequence in that chain determines what each protein will do, and also gives each protein a unique three-dimensional structure. We hear all the time about genetics and genes, but the actual function of the things is left vague. Here is their main function: They tell the amino acids how they will go together and make that protein. With prions, if that unique three-dimensional structure comes out wrong, a kind of hell is unleashed.

There are a number of diseases caused by prions, in both animals and humans. As noted, kuru and mad cow disease are among them. Scrapie, a disease of sheep, is a prion disease. In humans, so is Creutzfeldt–Jakob disease, a form of dementia that in its symptoms resembles Alzheimer's, but it kills more quickly. One that is thankfully rare is fatal familial insomnia, a strange disease, in this case inherited, in which progressive brain damage causes the victim to become unable to sleep. A wonderful book on this topic, one that explains prion diseases by telling the story of a family in Italy that has been cursed by fatal familial insomnia for over two centuries, is *The Family That Couldn't Sleep: A Medical Mystery*, by D. T. Max. He begins by clarifying how proteins are supposed to work, and what goes wrong with prions:

The body manufactures proteins as ribbons, which then fold into three-dimensional shapes that allow them to fulfill whatever function they have. Most proteins, because of the arrangement of their atoms, have only one shape, but prion proteins—the theory goes—have two, one normal, the second infectious and lethal.

Once the first abnormal prion appears, it begins to spread in the body through a process called "conformational influence"—"conformation" meaning "shape." After one prion becomes malignant, the way it's folded allows it to bond with adjoining healthy prion proteins, and the bonding causes these prion proteins in turn to fold wrongly and bond with others, making them fold wrongly, too. The chain reaction extends from protein to protein, normal prion protein changing to lethal prion in a series of Jekyll-to-Hyde conversions. Cells that have the misformed prions in them sicken and die, for reasons we don't really understand, but the effect of all these misfolded proteins on the delicate cells of the sufferer's brain is overwhelming: the brain tissue of prion disease victims is full of gaps, areas where all the cells have died, as if an explosion had gone off.

That is why scientists came up with the word "spongiform" to describe those brains. They are full of holes, like a sponge.

Furthermore, the misfolded prion is all the more deadly for being a protein. These complex molecules are the stuff of life, but are not actually alive—and so cannot be killed:

It is enormously difficult to disinfect a prion. What kills viruses and bacteria barely affects them. Boiling will not disinfect them, nor will heat. You can't reliably "kill" a prion with radiation. You can't pour formaldehyde on it to render it harmless—in fact formaldehyde makes prions tougher. Not all bleaches can kill prions and those that can need to be highly concentrated. Prions bond to metal. They can be spread, for instance, when doctors reuse the electrodes planted in patients' brains for EEGs or by dental equipment. To be on the safe side now, some hospitals discard the tools they use to operate on or autopsy those with CJD [Creutzfeldt–Jakob disease] after a single use. Prions endure in soil too. After a flock of sheep in Iceland came down with scrapie, they were killed, and the land left unoccupied for ten years. At that point, farmers brought in a new flock, which picked up the disease. Researchers once opened up a container with the preserved brain of a person who had died of a prion disease twenty years earlier and injected the tissue into lab animals. The animals contracted the disease.

So you get it: This stuff is bad news. And as I said, it is in Greater Yellowstone.

And plenty of other places around North America. Another prion disease is chronic wasting disease (CWD). Journalists sometimes call it zombie deer disease, which is a name we should outgrow, because it is not especially descriptive (besides, it has the causality turned around: Zombies want to eat brains, but to get a prion disease, you have to eat brains *first*). Like the others, it is a transmissible spongiform encephalopathy; it attacks the brains of the victim, leaving them literally full of holes. The visible symptoms develop, as often with prion diseases, after a long incubation period, lasting eighteen to twenty-four months after exposure. Young animals do not, therefore, show symptoms. The first of these is difficulty in movement, and then weight loss. There follows a series of behavioral changes: The victim's interaction with other animals changes, and it will develop both a persistent drowsiness and an extreme excitability when handled by, say, a veterinarian (I am using the Merck veterinary manual as my source for all this). What both vets and physicians call "ataxia" also develops, which is a lack of control over various parts of the body. The head will droop, and will also begin to shake persistently. As the disease enters its final stages, the animal will stare for long periods, at nothing. It grinds its teeth. Often, the animal will now be visibly emaciated—and yet the disease can be advanced while it is still at or near a normal weight. Victims die of a variety of causes. Aspiration pneumonia—that is, pneumonia that results when the animal breathes in the contents of its mouth—is a common cause of death. When a vet tranquilizes the animal, it may die, to the vet's surprise. Cold weather will kill it. Mainly, the victim will be vulnerable to the threats that beset all animals, and that one with advanced CWD is in no condition to avoid. It is always fatal.

Where CWD came from is the subject of ongoing debate. It was first described, and named, in 1978, although it was observed, in Colorado mule deer, a decade earlier. It has been spreading around North America ever since. It is a disease especially of deer, and another source of ongoing debate is what animals can actually get it. As noted, bison do not—or, at least, it *appears* that they do not. According to Merck, "It is found in farmed or free-ranging populations of mule deer, white-tailed deer, red deer, and elk (wapiti) in 26 states (USA) and 3 Canadian provinces, with the most recent identification of CWD in captive red deer in Quebec, Canada. Rare cases of CWD in free-ranging moose have been diagnosed in Colorado and in Alberta, Canada." It has recently turned up elsewhere, for instance "in South Korea, where a few elk imported from Canada were CWD-infected. In 2016, CWD was diagnosed in wild reindeer in Norway, which marked the first finding of the disease in Europe and the first case in wild reindeer. Since then, CWD was discovered in several Norwegian red deer and moose, and in 2018 and 2019, single cases in moose were documented in Finland and Sweden." Other animals may get it, but hardly ever. It is almost exclusively a threat to members of the deer family.

The prions that cause CWD can be acquired by a new victim in a variety of ways. Because prions are so hard to destroy, this new victim can get it from waste matter, from forage or feed, or just from the ground. They can also get it from other animals in situations in which animals crowd together; winter feeding is thought to be a danger, since animals that would otherwise be inclined to spread out will instead press together, sometimes—during, say, a hard winter, when the animals may be desperately hungry—into what amounts to a mob. The good news is that multiple cycles of freezing and thawing does degrade prions, and since the deer are often up in the mountains, this one form of natural sanitization may offer some protection. Scientists are, however, still learning about CWD, and nearly everything I have said about it is the subject of ongoing debate.

Often raging debate, and that brings us to Yellowstone, where all controversies, it seems, are heated. CWD is certainly in Greater Yellowstone; in 2023, it was also confirmed, for the first time, to be in Yellowstone National Park itself. In October of that year, a mule deer buck died on the Promontory, the long landmass that separates two of the southern arms of Yellowstone Lake—the true middle of nowhere, actually. The Wyoming Game and Fish Department had captured this deer near Cody and fitted it with a radio collar as part of a different study. Such collars have what biologists call a mortality signal, a tone that goes off when the animal has been stationary for a period of hours, meaning it has either gone into hibernation, gotten rid of the collar, or died. The biologist nowadays gets an email saying the signal has gone off. When Wyoming Game and Fish found the carcass, with the help of the NPS, they took samples and tested them; WGFD has been monitoring Wyoming deer and elk for chronic wasting disease since 1997. And this deer had it.

The disease had surely been in the park for a while; it was first found in Greater Yellowstone in 2018. In that year, a car hit a mule deer in Grand Teton National Park, and when the carcass was tested, the deceased animal turned out, like the mule deer on the Promontory, to have had CWD (and although we have no way of knowing what was going on in that poor mule deer's mind, that is how CWD victims often die: They are so impaired that they do not have the strength or speed to avoid cars). Since then, the US Geological Survey has, at its Northern Rocky Mountain Science Center, been evaluating what CWD is up to in the Greater Yellowstone Ecosystem.

They are trying to answer a number of key questions. How important are salt licks, for instance? Ranchers and farmers put these out for their livestock, but in a place like the Northern Rockies, the deer and elk use them as much as the cattle do. CWD surely must be transmitted through wild deer and elk herds this way; the question is how much of a problem they represent, and what can be done about it. Another question is what effect predators will have on the herd. Yellowstone now has more predators than it did before, and it is hoped that they will cull the infected animals without getting sick themselves (wolves, cougars, and bears do not, it appears, get CWD); researchers associated with the USGS are looking into this matter, a more complex question than it appears because it takes so long for a prion disease to run its course in a single animal, in a local herd, and in entire populations.

Special concerns surround the National Elk Refuge, a peculiar area northeast of Jackson, Wyoming, which is managed by the US Fish and Wildlife Service. The exterior locations for the movie *Shane* were shot around the town of Kelly, Wyoming, which is on the north boundary of the wildlife refuge. It was established in 1912, on land purchased by the federal government from a homesteader—ironic given that *Shane* is about homesteaders run off their property by an evil boss. It was intended mainly as a place for the elk to gather in the winter, and today it is also a place for tourists to gather in the winter. Special sleigh rides run out into the refuge from Jackson during the snowy months. Elk press onto the refuge in great masses during the winter, where they are given supplementary feed, and the grass on the refuge is further irrigated to grow forage for the animals. The money, by the way, comes from a longstanding arrangement whereby the Boy Scouts gather elk antlers shed by the bulls and sell them, keeping part of the money and putting the rest into the irrigation fund; as I said, it is a peculiar place. Given that they are crowded together like that, every winter, it is difficult to imagine that CWD is *not* moving through those elk, and it is worth noting that it first appeared in Greater Yellowstone with that mule deer hit by the car in Grand Teton National Park, right next door. But it is not clear what impact CWD is going to have; at this point, what we have are educated guesses.

The US Geological Survey is coordinating research into the National Elk Refuge feeding and crowding issue, and in a separate study, is looking at alternatives to winter feeding. They are examining the animal side of CWD closely. The other side is the human one, and it may be the most contentious of all.

Can humans get CWD? Again, the evidence is subject to ongoing debate, and the debate here is especially heated. Much is at stake. States with large populations of elk, mule deer, and other members of the deer family—like Montana, Wyoming, and Idaho—get a great deal of their state gross domestic income from hunters, and even people who do not earn their living in industries associated with that hunting still eat a lot of wild venison, generally killed and dressed by their own hand. My neighbors in Gardiner eat wild elk and deer meat (and moose, antelope, bighorn sheep, etc.) in such quantities that it might be measured by the ton. Where, then—they ask—are the human CWD victims? There have been none, and the hunters do not appreciate the suggestion that their traditional lifeways are irresponsibly dangerous. But we are very early in the era of chronic wasting disease in Yellowstone, or indeed anywhere, and the jury is still out.

If it does happen, it will be rare, and the evidence points both ways, although all the research supports the obvious claim that the risk is low. Studies have looked to see if human prion diseases like Creutzfeldt-Jacob afflicted more people in places where the deer herds have CWD, and the results were negative. Similarly negative were studies that looked at what may be the best test animal of all: hunters who eat a lot of venison. Research continues to appear that suggests CWD could claim human victims. One report published in 2022 used humanized mice, animals that were modified to have genes, cells, tissues, and sometimes organs from humans. Creepy stuff—look into this matter, and you are in for some weird science. The researchers injected CWD prions into these humanized mice, and they developed CWD, and even shed prions in their droppings. Still, the most recent study to appear as I am writing this found the opposite. In June 2024, the CDC journal *Emerging Infectious Diseases* reported a study in which researchers grew cerebral organoids, artificial tissue that resembles part of the human brain. When they exposed the organoids to CWD, the exposure did not take. The organoids remained free of misfolded prions.

If CWD were a threat to humans, we would have a positive answer by now. Instead, the risk, it is clear, remains exceedingly low. Nor are the prions involved going to have their DNA mutate, turning them into a ferocious killer. They have no DNA. One defining feature of misfolded prions is that they stay the same, decade after decade. That is where we are at present, and that is where we are likely to stay. The threat is to the deer and elk, which is bad enough.

Here is an odd angle to all this, an angle that just now occurs to me: There exists a genus of lichen that may lead to a treatment for prion diseases, or at least lead in that direction, and we have that lichen in Yellowstone. It is reindeer lichen, of the genus *Cladonia*, in various species. An ecologist at Montana State University, Bozeman, Sharon Eversman, surveyed the lichens of Yellowstone National Park in the 2000s, and published her results in *Yellowstone Science* in 2007. Reindeer lichen got that name because they are a food for caribou and reindeer, in the far north, but we do have them in the park: "'Reindeer lichen,' species of *Cladonia*, are rare in the Yellowstone region," Eversman wrote, "but stunted forms of *Cladonia mitis* (gray-white, abundantly branched) can be seen in three thermal basins—Biscuit Basin, Norris Geyser Basin, and Phantom Fumarole on the Pitchstone Plateau. Reindeer lichens are more common in moister areas north of YNP," and it seems that they are in the geyser basins because it is wet enough for them there. *Cladonia* has an enzyme in it called serine protease, and it has been shown to degrade prions. It may prove useful in future medications against prion diseases.

So, if you find yourself turning into a zombie, you might eat the lichen. Just stay away from the wolf lichen. It will make you sick as a dog.

Chapter Seven

A Different Kind of Bullfight

The Bison

I first saw Greater Yellowstone in the late morning and early afternoon of May 21, 1990, and first entered Yellowstone National Park at 11:30 a.m. on that day. I do not have to look at anything like a diary to get that date, or the time, and in that era nothing had a time stamp on it anyway (cameras required film, and phones were in phone booths, where they belonged). I remember because that day, that week, that summer changed everything for me, for the better, indeed for the best.

Two hours later, after lunch at the old West Thumb Hamilton Store, now gone, I was in Hayden Valley, the wide, rolling, open landscape north of Yellowstone Lake and south of Canyon, when I passed another milestone. The "stone" in that last word is appropriate, because what I saw was a living creature, but . . .

. . . but it was like he was part of the *geology*.

I had read widely and not too deeply before I left home in Burbank to drive here. My starting point was a state of almost complete ignorance. When a friend had told me about jobs in national parks, and said that she worked at a park in the mountains that I remembered later as beginning with a Y, I assumed, naturally, she was talking about Yosemite. I only learned the difference at about the time I was hired to work at the job she and her husband had arranged for me, where they in turn had worked in the 1980s (they had met here in Yellowstone and fallen in love, a common outcome). Once I knew where I was coming, I launched a "research" project on this Yellowstone place, which I now understood was the park with the geysers.

It makes me smile now to look back on, but I was deeply bewildered at the time. In a used bookstore (there were still at least a half dozen of those in Burbank alone, and a Borders and B. Dalton close by), I found a guide to the geysers that was intended for the one-day tourist in a great hurry. The photos of the thermal areas looked, to me, when I got the book home and opened it, like a pastry shop hit by an improvised explosive device, and after an evening flipping through it, I put it aside. As I write, my guide to the geysers sits in a box under the bed, down the hall, and it is like the mummy's tomb, unopened in all these many years—cursed, some say. I found a more serious book about the forests and volcanism that made me feel like I would be living inside the movie *King Kong*, in the part on the jungle island. None of it made sense. I gave up on all books, and decided I would just have to learn from the real thing, which was actually a good decision, especially given the age of some of my sources. Remember: no internet, in this world we inhabited. I had somehow contrived to find a magazine article about the animals that itself referred to sources from the 1890s. The author meant no harm, but with sources that dated, the article left me with a mental image of the park that was full of holes. I came away believing, for instance, that there were only a few dozen bison left alive in North America. Would they even survive long enough for me to see one, or would they go extinct first? I was angry about it. I certainly never expected to see one in the course of just one summer.

So it is that I have always been patient with park visitors who slam on the brakes and swerve drunkenly and stop in traffic when they notice an animal beside the road in Yellowstone National Park. Because that was what I did when I saw my first bison.

There were, at that date, about three thousand inside the park.

I actually sat there and stared. My car was a 1970 Chevy Chevelle station wagon, more like a construction project than an automobile. I may have blocked the whole road. It was, however, so early in the season that when I finally woke up, there was no one behind me. I was able to sit there and study him, my mouth, I suspect, actually open.

This animal was a full-grown bull, two thousand pounds of bad attitude nibbling, at that moment, on dandelions in the roadside ditch. I did not even know he was a bull at the time; growing up in Los Angeles had left me good at reading gang graffiti, and bad at any kind of barnyard skills. Telling a bull from a cow (the bison young are, again, calves) is easy; in any given group, the adult females are visibly smaller. Even more useful is the head and horns. A female has a narrow head, and the horns are smaller. As the cow grows older, the points of the horns will curve back toward the head. The bull's horns continue growing straight up, and the horn is larger. A standard way to make the comparison is that a cow's horn is the same diameter as her eye, while a bull's is twice the diameter, though that is not a good way to spot the difference, because the eye is dark brown against a dark brown background, and if you are close enough to make that comparison, you are in some danger. The bull has more hair atop his head, sometimes even wearing a kind of Zoomer broccoli haircut—but there all resemblance with the human body ceases. His head is otherwise distinctive for its astounding size, with a wide forehead that, when you are as close as I was in the car, is the business end of a medieval battering ram come to life.

A bull bison—full grown, profoundly dangerous.

Here is one feature of this experience—such a common one, during even a brief visit to the park—that made it breathtaking: I was closer than I could get on foot unless suicidal. I was in a car; it was both legal and sane, and so this monster was *right there*. I did not have a good sense of perspective, sitting behind the wheel on a bench seat with rotten foam rubber inside it, but it looked like the animal was actually *above* me. I was correct. He would have been about six feet tall at the top of the alpine hump on his back, and maybe even taller. He might have been as much as eleven feet long.

He at last moved along. I lifted my foot off the brake and rolled along with him. His head bobbed like the head on an oil field pumpjack. He nibbled another dandelion, another, another, and at length, moved by an impulse that will forever be beyond our understanding, he turned away from the road. Animals will frequent roadside ditches because the vegetation is green there late in the year, but it was spring here, with snow in the shaded folds of the valley and ice in the coves on Yellowstone Lake, creaking and snapping in the wind. It was wonderfully early in my summer, green with promise. Hayden Valley rolled away from us, looking like Ireland in its uniform springtime emerald.

The bull climbed out of the ditch, and accelerated as he did so, leaving me with one last improbability: He broke into what was almost a trot. It ought not to have been possible to move that mass in such a spritely way, but of course he made it look easy; you will hear it said that they can jump over an object five feet off the ground, and they can run thirty-five miles per hour. I have seen them do that many times. I at last looked in the rearview. Another vehicle was approaching. I got on with my life, with pleasure, as it was taking such a miraculous turn.

I looked at the bull one last time. He was in the prime of life, and therefore was invincible. At his age, only humans could threaten him.

And he could threaten them right back. Human technology alone made him vulnerable, when we are around.

He is dead now, I only just realized (so, by the way, is the car). As indestructible as he looked that day, he surely is, because while you will see various estimates, bison live between fifteen and twenty years in the wild, and typically twenty-five in captivity. The world record is held by a bull who died at thirty in Golden Gate Park in San Francisco. I would rather not know how he died, especially if it was by a bullet. Unlike his Hayden Valley mates that day, I will never forget him. I am doing what I can, here, to make him a little bit immortal.

And why not? The main lesson I took from him, that day, was that this creature, officially *Bison bison*, is full of surprises, so much so that I might be forgiven for making him some kind of spirit animal, if I were into that kind of thing (my girlfriends in the park somehow always were). Life in Yellowstone would develop into one long surprise for me. Maybe he could have sprouted wings and shown me the way.

In the background is the Old Faithful Inn. It has played host to bison probably since it was built.

The bison and Yellowstone National Park go together. The park was not made to protect them; it was made to protect the geysers and the canyon. They were what riveted the first more-or-less official exploring parties, and they were what moved Congress to make this strange gesture. By throwing a cordon around the thermal features, however, the lawmakers also created a kind of legal corral that would, over time, become a wildlife refuge, too. And while the bison was slaughtered elsewhere in North America, here, in a bastion made partly by law and partly by geology and weather, they hung on—if, for a time, just barely.

If the Yellowstone bison herd had gone extinct, the species itself could well have disappeared. Human-caused extinction is almost always a calamity (smallpox and polio maybe could go), but the loss of the bison would have been unusually tragic, because thousands of years of human life on this continent have been intimately bound up with this creature.

Consider the epic tale of the Sioux. Members of this tribe in the present day live on reservations, especially in Montana and the Dakotas, but they once lived in the forests of what is now Minnesota. They

Mother and child, in the Lamar River valley.

essentially fought their way onto the Great Plains, displacing the tribes in their way as they went, and adopting the horse and gun as these items came their way. Thus equipped, they became the fearsome warrior tribe of the frontier West. Nathaniel Philbrick, in his book about the Battle of the Little Bighorn—fought by a native warband of, predominantly, Sioux—explained the process. As the Sioux moved west toward the Missouri River,

> *they came to depend on the buffalo as the mainstay of their way of life. When the French explorer Pierre Radisson met the Sioux in 1662 he described them as "The Nation of the Beef."*
>
> *By the middle of the eighteenth century, a combination of events had set the stage for the rise of the western, or Teton, Sioux. Being a nomadic people, they were less affected by the diseases that began to devastate their more sedentary rivals along the Missouri River. The gradual acquisition of firearms made the Sioux an increasingly formidable foe, but it was the horse, obtained in trade from tribes to the south, that catapulted them into becoming what one scholar has termed "hyper-Indians."*

It was all driven by calories from *Bison bison*, and ultimately from the grass of the Great Plains.

The people of the frontier called the bison "buffalo," of course, and they are the same animal (the real buffalo of the Old World, of Africa and Asia, do not actually much resemble our buffalo; Vietnam veterans will know the Asian kind from the rice paddies of Southeast Asia). During the nineteenth century, the bison of the North American plains were amassed into great herds that have passed into legend. Evan S. Connell describes them eloquently in his book *Son of the Morning Star*. The men he names were soldiers or explorers who were eyewitnesses to the reality of these great masses of bison:

> *Two herds, each so vast that no reasonable estimate was possible, had darkened the plains, one above and one below the Platte [River]. Frémont, who traveled through this region in 1842, found himself*

surrounded—the herd extending for several miles behind him and forward as far as he could see.

Francis Parkman saw them in 1846. Working on a history of LaSalle he reflected that the French explorer, too, must have observed a wondrous spectacle, ". . . the memory of which can quicken the pulse and stir the blood after the lapse of years: far and near, the prairie was alive with buffalo; now like black specks dotting the distant swells; now trampling by in ponderous columns or filing in long lines, morning, noon, and night, to drink at the river—wading, plunging, and snorting in the water—climbing the muddy shores and staring with wild eyes at the passing canoes."

Twenty-five years later not much had changed. Col. Dodge notes that during spring migration the buffalo sometimes would move north in a single column perhaps fifty miles wide, of unknown length. At other times they traveled in parallel columns, all marching at the same pace, blackening the earth. On one occasion when he was engulfed by a migrating herd he climbed Pawnee Rock to get out of the way and saw the prairie covered with buffalo for ten miles in every direction.

The Sioux made them the center of their universe:

Pte, *the buffalo, provided just about everything they needed, right down to his tail which made a splendid fly whisk. Fresh meat, tallow, warm robes, leggings, bow strings, bone needles, battle shields and coracles made from his tough hide, axes and hoes from his shoulder blades, sledge runners from his ribs, glue from his boiled hooves, red paint from his blood, fuel from his dung, ladles from his horns, hair to stuff pillows, and so on. They even used his long black beard to ornament their clothes. Therefore they addressed him as Uncle, this useful monster, and followed him across the seasons.*

An advantage of spending many nights in Yellowstone is that my friends and I have had occasion to burn bison droppings, the "buffalo chips" of the Old West. Because there was so little wood on the plains, settlers and natives both used them as fuel. They do not burn like wood. The fire is

A Different Kind of Bullfight

A bull on the run. GETTY IMAGES

more like a coal or peat fire, glowing rather than flaming, and odorless if the material is entirely dry. On one occasion, we got a smell like burning hair. They certainly do not smell as human sewage would, if you contrived somehow to burn it.

We have, in Greater Yellowstone, some of the mementos of this life. If you drive out of Yellowstone National Park and head north from Gardiner, a little over twelve miles away on US 89 is a point on the Yellowstone River called the Yankee Jim River Access. You will now be entering Yankee Jim Canyon. As you do, you may notice your vehicle passing over a bump, more pronounced during the winter. As bison migrate down out of the park during the cold months, they are stopped here by the mountains and river; the bump is a cattle guard that keeps them from walking down the road. This barrier separates them from the cattle herds down the canyon, in case the bison might be carrying brucellosis—more on that later.

US 89 thus becomes a kind of bison historic route, linking the new with the old. Beyond Yankee Jim Canyon, the rugged slopes fall away from the road, and you enter Paradise Valley. The glaciers went this way, grinding across the landscape in roughly the same direction the Yellowstone is flowing, and that your vehicle will be rolling; you cannot see or perceive it, but the route north runs gradually downhill. On either side, look at the top of the ridgelines. Notice how uniform they are. That is a volcanic lava flow from some millions of years ago. Out the right-side window (to the east) is a landform known locally as Hepburn's Mesa, after a wonderful local eccentric who was an amateur paleontologist and ran a museum at the base of the mesa in which he showed off the fossils he had found here. The lava flow atop the mesa is 2.2 million years old; I don't know about the others, and what I have learned of geology has taught me that the situation is always radically more complicated than it appears, so we will call them "lava flows" and leave it at that. The native people in the area looked at the geology differently, and they knew a good thing when they saw it. They used these cliffs for thousands of years as the crucial step in what was one of their favorite ways to hunt.

Here were a number of buffalo jumps, apparently a large number. I got to see one once in the company of some friends and a local cowboy who grew up learning about the ancient natives in a hands-on way. The jump we saw was to the north, next to the town of Emigrant, but I know of at least two on Hepburn's Mesa, and our cowboy thinks that there are hundreds in the valley, some obscured enough by the passage of time that they would be difficult to recognize. The one he showed us was, after he pointed it out, blindingly obvious.

But only after he pointed it out. We had talked him into revealing his secrets, the secrets of the landscape, hiding out in the open. We parked on private land belonging to one of his neighbors, and he led us to a spot where the cliff is close to the valley floor. I was surprised: It was not the sheer plummet I had expected, hundreds of feet tall. "That fall killed them?" I asked.

"I don't think killing was the idea," he said, "or at least not until later. They ran off the edge there, and then bounced down on the rocks, and were in pretty bad shape by the time they were done."

He gestured as he spoke. We were at the base of the lava, here shaped into the columns that give "columnar basalt" flows their name. The lava was the same near-black shade it normally is, amid a landscape of high Montana desert. We were on a working ranch; a cow path passed along the bottom edge of the lava. The basalt had, over the millennia, fractured and toppled downward to form a jumble of blocks. We could almost have climbed to the top of the cliff up those blocks, except that they looked terribly unfriendly, covered with jagged edges like the ribs and spine of a starving creature, with—we guessed, not willing to check—prickly pear cactus growing in between. Along the top was a spot where the cliff really was sheer for six or eight vertical feet. That is where the bison were launched on their terminal trajectory.

"Everything came together for them here," the cowboy told us. "The wind is from the right direction. The cliff is the right height and shape. And there are all these cobbles."

He pointed now at the rocks lying here and there, ground into the size and rough dimensions of a softball, then dropped here by the glaciers.

"That over there is quartzite, I think," he continued, pointing to another, smaller rock. "The bigger ones are just the thing for finishing off the bison. Then you use one cobble as a hammerstone on another, and make blades to butcher your kill."

The basalt cliff. The ancient people drove bison off this cliff for hundreds and maybe thousands of years.

"Must have been a messy business," I ventured. What had happened was clear now. The ancestors of the Crow, the Shoshone, and other tribes had driven bison off the edge of the cliff, above our head. The panicked animals had been propelled by their own momentum out into space, then rapidly down, bellowing, flailing, onto the jumble of lava below, where their great bulk now helped kill them. Rendered powerless by their broken bones, they were finished off by the half of the hunting party that remained below while the other half chased the bison off the cliff.

They were, we now saw, butchered on the spot. "How did you learn about this place?" one of us asked.

"You learn about them eventually just growing up here," he said. "But look at the ground."

We did. It was . . . not right. Being originally city people, we thought it looked like cat litter. I got down on one knee, and saw that it was . . .

"Bone," the cowboy confirmed. "The settlers used to dig in places like this and use the soil for fertilizer, there was so much bone in it."

I recalled now that the bone was the giveaway, the announcement, if you were looking for one, that the cliff above had been used as a jump. I had expected bones like a dog would chew and bury. The chips were a half inch long, a quarter inch, a sixteenth, and smaller, and the ground all around was off-white with it. Some had been burned. Maybe the camp had been here, we said, speculating; there was a creek close by, another advantage to this site. It was easy to imagine a wild feast right here, but we have no witnesses to the actual operation of this jump, no one who ever wrote down the details. What we have is the archaeology.

We took an easier route up onto the bench above, and now saw how the whole arrangement had worked. The artificiality was suddenly obvious. Extending back from the cliff were long lines of individual piles of cobble, what archaeologists specifically call cairns when humans build the piles, as they had here. "Brown says that the way it worked," I volunteered, "was that they built these cairns in two rows and linked them with wood and lariats . . ."

"And hung feathers from them," our guide continued. He and I had seen the same literature. I had specifically found an article published in the 1930s by Barnum Brown, a paleontologist who examined a bison

The marks of the hunters' passage: millions of bits of charred bone.

jump site in this exact part of the world. If you know a nine-year-old boy who is into dinosaurs, he can tell you who Barnum Brown was: In 1902, he found the first *Tyrannosaurus rex* known to science, as only one small part of a long career (as they often do, the tyrannosaur turned up here in Montana). The bison jump he documented was another small part; it is hard to work out, but it may have been the same one we were now looking at, as the sites, in this part of the world, tend to blur together.

Our guide gestured along the line made by the cairns, which extended far enough almost exactly due south that they faded from view behind the distant sagebrush. The wind was gusting the opposite way, almost exactly due north, as it normally does here, blowing in the direction the Yellowstone River flows. The cowboy naturally knew more about bovines than any of us, and what he knew about the kind of cows that live in Paradise Valley now informed what he could make out about the way the jump had functioned long ago. "People think they're stupid," he said. "They're not. They just see the world differently than we do."

He gestured again, pointing toward the far end of the line of cairns and running his finger south to north to indicate the line of travel. "The wind was blowing downvalley, the same way it always does," he continued. "The hunters got upwind of the bison. The smell of humans would have gotten the herd moving down the chute. When they're running behind and in front of one another, they don't see the same way we do. They see the animals all around them." A bison's eyes are arranged as prey animals' eyes normally are: They focus not on what is in front, as we do, but all around. My veterinary students learn it like this: "Eyes in the front? Likes to hunt. / Eyes on the side? Likes to hide." An herbivore's vision is panoramic. We are naturally impressed by their astounding bulk; however, the bison is still, like the deer and the antelope, an herbivore. When threatened—here, by the odor of humans on the wind—they run away.

"They'd be pushed and pulled by the herd, both pushed and pulled," our guide continued. We walked now between the last of the cairns, toward the brink. "The wind made them move away from the smell of humans back there somewhere, and they wouldn't see the plunge until it was too late. The animals at the front would see the danger, but . . ."

We arrived now at the edge of the basalt, where it made a near-perfect right angle. We stood with our toes on the edge of the brink—and then backed away. The jumble of broken lava below was horrible to contemplate.

"They would see the danger," he finished, "but it would be too late."

The animals in front would be pushed over both by the group and by their own weighty momentum.

Over many generations, the native groups that lived together with the bison learned to step inside the animals' consciousness and see things the way they do. Today, as we looked the whole site over, the big remaining mystery was how the cairns functioned. Each cairn, it seemed, had an upright post in the center; the posts were joined by lariats—ropes of woven leather, presumably bison leather—and the whole made more impressive, somehow, by feathers inserted into the lariats. The fluttering of the feathers, in the considerable wind that has always regularly barreled down this valley, must have looked unnatural and threatening to the bison, enough that they held to this fatal course, down the middle of the two lines of cairns. The barriers were otherwise insubstantial, like the crowd control stanchions you see in places like amusement parks, where the stanchions guide people as they line up for the rides. Would barriers that flimsy actually line these big animals up and keep them all moving in a row in the same direction? Well—it works on humans, doesn't it? Disneyland depends on it.

A day like this, on the ground and actually seeing and touching the rock, brought the distant past to life, brightly. I had seen artists' renderings of jump sites like this one, and they had led me badly astray. One, I recall, showed some cavemen standing serenely next to a sheer cliff a hundred feet high, down which chubby wild oxen plunged to an instant death. No, the animals here were badly wounded by the fall, and the people at the bottom then had the treacherous task of actually killing them, which would have been routinely terrifying and maybe even deadly. I had thought that this technique would be wasteful; I was envisioning tens of thousands of animals dying, but judging from the scale here, it was more like dozens. Not all speculation about these people is equally uncertain. Here is one statement we can make and be certain about: The hunter-gatherer life that they led, for millennia, was not easy. If you got hungry in the middle of the night, there was no convenience store full of calories down the block. What happened here after that last bison was dead must have been a feast of a lifetime, a feast you tell stories about years later. Surely, they were not in the mood to be frugal.

The cairns had one other message to impart: They were splotched with the slow-growing green lichen that showed they had been piled this way for well over a hundred years. Had it been two hundred? All we had were guesses, although our speculation did run up against one hard limit in time. That limit was the year 1884. The hunt would be impossible after that date, because that year, the last bison hides were shipped from what became the state of Montana.

The continental herd of bison—the great mass that explorers encountered when they pushed west in the nineteenth century—was so large that today our estimates are not reliable to the nearest million, and maybe the nearest ten million. One relatively conservative estimate places the total headcount at twenty-five million. After the explorers and the first settlers came the railroads. In our part of the world, the Northern Pacific Railroad took a logical route along the Yellowstone River. At Livingston, a branch line headed south, following the river to the now-extinct town of Cinnabar; the official first train arrived there on September 1, 1883. Eventually, three more miles of track brought the trains all the way to a station at Gardiner, next to the Roosevelt Arch.

The railroad is gone now. Scheduled service ended in 1948, and the last train to visit Gardiner was a charter in 1955. The automobile's reign had begun. You can follow the bed where the tracks ran almost all the way from Livingston to Gardiner; the private property lines are the major barrier. Visiting the bison jump, we saw that the rancher who owned the land was using Northern Pacific rail ties as fence posts. They were close to indestructible.

The railroad from Livingston to Cinnabar was intended mainly to carry tourists. The railroads in general, however, also carried men with guns.

So the rails brought the slaughter. Market hunters shot bison by the thousands. The herds on the central plains went first, because the railroad got there first, providing a way to ship the marketable parts of the dead. Montana's bison lasted longer; there were herds with hundreds of thousands of animals in them well into the 1880s. What made bison worth shooting was their hides, which were stretched and dried and shipped in piles. The hunters regularly left the rest. It happened even as the bison entered the culture as a noble symbol of the western frontier.

Transporting the hides to East Coast markets required technology, and steam power was the answer even before the railroad: The Yellowstone River became a highway that steamboats used to haul hides eastward before the trains made it this far. The local newspapers of that time record the mind-bending numbers. In 1881, for instance, the Bozeman *Avant Courier* reported that one hundred thousand hides would leave Montana Territory that year. As big as the herds were, they went quickly, and by the mid-1880s, there were no more.

Except.

A Different Kind of Bullfight

In the high mountains to the south, there was a hideaway: Yellowstone National Park. Almost unnoticed, the last wild bison—without knowing they were doing so—holed up and waited for a better day.

And it came—but not until they had flirted with extinction. Mary Meagher, the government's Yellowstone bison expert for many years, looked through the archives and determined that the low reached twenty-three individuals in 1902.

It was touch and go for a remarkably long time, and in 1894 a great villain entered the picture to push them closer to the brink. During the late winter of that year, we don't know exactly when, a sometime resident of Cooke City, Montana, named Ed Howell left that town on skis, pulling a toboggan and accompanied only by his dog. Howell evaded the US Army guards who covered that part of the park; during this time, Yellowstone National Park was still patrolled by the army, and they were stretched too thin to guard against this kind of intrusion. Howell skied up over Specimen Ridge and down to his camp on Astringent Creek, a tributary of Pelican Creek, which the modern East Entrance Road crosses north of Yellowstone Lake. He had been to Cooke City for supplies, and had been living in this primitive camp since September. It was a praiseworthy feat, chugging all that way on nineteenth-century skis, which might be twelve feet long, and made, naturally, of not-at-all-lightweight wood. He was not, however, here on a praiseworthy mission.

Happily, another man was. Felix Burgess was a tough old frontiersman who was working as a civilian scout for the army. He was out patrolling with two unarmed companions when they found a tree with a half-dozen bison heads suspended in its branches. The heads were hanging there to keep scavengers from getting them. Burgess was actually not surprised to discover this, the ugliest parody of a Christmas tree ever to grace the park. He might not even have been surprised when the three heard rifle fire. Six shots echoed through the wintertime forest, and Burgess knew he was in business. He knew who Howell was, and he had suspected he might find him here. Howell was, during this time, the park's most accomplished poacher. He was engaged in his chosen trade, shooting bison and beheading them. Taxidermists would pay hundreds of dollars for a single head; the animals were now so rare that scarcity drove the price up until it is a wonder that Howell alone was shooting them.

Burgess found Howell in the distance, cutting up a freshly dead bison. He had killed five, missing with only one shot, and so had now killed a total of eleven. Burgess was armed only with a .38, but the wind was blowing in his face, masking his scent from the dog, and also masking any noise he made from Howell. Burgess managed to cover the ground ("I expect probably I was pretty lucky," he said later), and Howell knew nothing until he looked up to see the muzzle of Burgess's .38 pointed at his nose.

Burgess and his companions hauled Howell and the dog to the Lake Hotel, where Burgess used the phone—the park had a telephone system now—to call in to headquarters at Mammoth. He made his report to the superintendent of the park, a US Army captain named George Anderson. One of the best of the army superintendents, Anderson was playing host this night to a newspaperman, Emerson Hough, who used the army telegraph to send the story to his employer. That employer was no ordinary newspaper; it was *Forest and Stream*, an outdoors journal edited by George Bird Grinnell, who had turned it into an influential voice in what we would call environmentalism, and Grinnell would probably call "preservation." I have told this story at greater length in two books, *Rough Trip through Yellowstone* and *Yellowstone's Lost Legend*, so to be brief, let me just say that at this moment, the Yellowstone bison, perhaps all bison, may have been saved.

We cannot know, but they would have had a rough time otherwise. The amazing fact about this affair is that Howell was not guilty of any serious crime. There were no federal laws in place to protect the animals of the park. Howell would sit in the prison at Mammoth, uncomfortably, but in a week or so, the Secretary of the Interior, in line with the standard procedure, would order Anderson to turn Howell loose. For violating the US Army's regulations, he would lose his rifle and be kicked out of the park, but his major regret would be the loss of those heads. He would not otherwise be inconvenienced.

That changed now. Hough telegraphed Grinnell, and the next day, members of the House of Representatives heard the story, too. With remarkable speed, a new federal law, the Lacey Act, was on the books. From then on, poaching animals in any national park would be a serious crime, one punishable by a fine of $1,000, two years in a federal penitentiary, or both.

Hough was actually in Yellowstone for another purpose; luck and Providence stationed him in Anderson's parlor in time to make this crucial contribution. He had been sent by Grinnell to ski through the park and count the animals. Led by the redoubtable guide Billy Hofer, he did so. They found plenty of elk, but where the bison should have been, they found mostly an ominous emptiness. When they were done, Hough reported the numbers in *Forest and Stream*: "We were now well settled in the conviction that the number of buffalo left alive in the Park was not one-half or one-third that generally supposed," he wrote, "and from what we had seen we feared that the killing had been heavier than anyone had dreamed. I would state here that I think our view of the case was conservative and fairly accurate. . . . Let us wait till some one has seen in one day and in one herd 200 buffalo in the Park before we ever again believe there are so many as that left. I do not believe there are 150."

And that number would continue to sink. A census in 1901 found twenty-five of them, and Congress then coughed up $15,000 for the park to use to purchase twenty-one more that were in the hands of private ranchers. They were kept, along with animals from the wild herd, in what became the Lamar Buffalo Ranch. Over the years, a series of building projects created the corral, barn, and "buffalo keeper's house" at the ranch. Today, up and down the river valley, the bison roam. If you see a single bison next to the road while driving elsewhere, don't do what I did, slamming on your brakes and sitting to look at that one animal. Go to the Lamar River valley, where there are regularly hundreds in view at once, and where you can honestly get a sense of what the West was like when the big herds still existed.

The animals there now are descendants of the animals bred at the Lamar Buffalo Ranch, which also sent bison to repopulate other ranges. So, *Bison bison* was saved in the nick of time, a good thing for more than one reason. For a country that is full of towns and cities and sports teams named "Buffalo" to have no buffalo in it would have been embarrassing at the least.

They are serenity itself, standing there like that bison, my first, in Hayden Valley long ago. As herbivores do, they spend most of their time eating. They stand, they drop their heads, they take a bite, they chew, and they do it again and again and again (this is fun to watch in the winter, when they use those monstrously large heads to bash the snow aside). Their attitude toward time is not ours. They have all the time in the world, it seems, to chew and bite and chew. Serene.

And that is how they unwittingly tempt humans into trouble. They do not look dangerous, and yet they are, I have always thought, the most dangerous animals in Greater Yellowstone. The statistics seem to confirm the claim.

Lee Whittlesey, in *Death in Yellowstone*, does not disagree, and he has an excellent explanation of why people are willing to approach these beasts close enough to get into trouble:

> *The bison, or buffalo, is a mythologized animal. To many Yellowstone Park visitors, it is not a genuine living, breathing creature. Instead, it is a painting, a symbol of a vanished past, a vignette of nineteenth-century America, but certainly nothing real. Many visitors want to approach it, to touch it, to somehow establish a close link with it, as if that might somehow connect them to their own frontier heritage. Having been a Yellowstone resident for more than thirty-five years, I sometimes lose sight of that simple truth. To me the animal is very real and very dangerous. But perhaps that mythologized perception of the animal is part of the reason that visitors have been injured and killed by bison in Yellowstone National Park.*

That is nicely put. You could object that it is speculative—but speculation is required to explain the lunatic persistence with which people continue to walk right up to this armored fighting vehicle of an animal, expecting that no harm will come of it. This is an animal that is documented as having, in 1890, attacked a stagecoach. I have seen them attack automobiles, and Whittlesey reports that he personally has seen them attack tracked snowcoaches. Yet people still want to make friends.

No one seems to know who had the honor of being the first person hurt by a bison in Yellowstone National Park. No doubt it happened during the trackless millennia between the retreat of the glaciers and the arrival of the first white explorers. It probably happened regularly. Having had a close look at that bison jump, I am certain that people were injured and killed in that exact activity, by bison who were hurt, but not so badly that they were incapable of having their revenge. What happened there, though, remains a matter of educated guesswork.

A good place to start the modern history of bison-related violence is with Rocky Mountain Dick, a classic Yellowstone eccentric whose legal name was Richard W. Rock. He had a kind of wild game ranch near Henry's Lake, not far outside the West Entrance to the park. Here, he kept and bred wild animals that he had caught in the surrounding wilderness—including, according to accusations leveled against him, the park, which was of course a form of poaching. Running a ranch for wild animals was not as odd a thing to do as it may sound; Rock made plenty of money hauling animals to the Northern Pacific station in Bozeman and shipping them to wealthy collectors and zoos. A reporter who visited him in 1901 wrote, in the *Hancock Democrat*, that he had "fifty-two buffalo, three grizzly bears, sixty elk and large numbers of moose, deer and Rocky Mountain sheep and several black and brown bears." Wearing a mustache, goatee, long flowing hair, and wide Mexican sombrero, Rock had something of the showman about him, and he delighted in using the animals for improbable purposes. He was photographed using one of his moose, the one he named Nellie Bly, as a carriage horse. His favorite trick, the one that he at last tried one too many times, was to climb on the back of a full-grown bull bison and ride it like a mare.

He explained the ways of the bison for another reporter, this one from *The Wide World Magazine*:

> "At close quarters, when enraged, the buffalo is a dangerous customer. He can deal a nasty cut with his short, wicked horns; and with head lowered as a ram he can land with terrific force, in regular catapult

fashion, a blow from which neither man nor beast could ever recover. Once lassoed and thrown on his back, however, he is easily managed and can be readily handled. In captivity, buffaloes usually prove very tractable, and can even be made companionable by kind treatment." In proof of this statement, Mr. Rock proceeded to leap on the back of a huge fellow, which carried him around the yard several times, without manifesting the slightest uneasiness or resentment.

Perhaps the resentment was building.

His friends tried to talk him out of doing it. One of them, a man named Kirby Garner, had warned him several times: "Dick, that buffalo will kill you someday. You had better be careful." Garner was referring specifically to a bull bison that Rock had had since he was a calf, and that Rock had named Lindsay. On March 22, 1902, the inevitable happened. Rock was in the corral feeding Lindsay when he turned his back on the animal. Lindsay was nine years old and in the prime of life; in photographs, he is every inch the bull bison. Rock was six feet four inches tall, but Lindsay made him look small. Lindsay attacked.

Lindsay first pinned Rock against the corral fence, and then went to work with his horns. What Lindsay delivered has, in the bullfighting rings of Spain, a name—*cornada*, a horn wound, the impalement that is generally how matadors die when they are killed in a bullfight. It is like being run through with a sword, only this sword is as wide as a bull's horn. Lindsay delivered a total of twenty-nine such wounds. He caught Rock on a horn, picked him up, threw him through the air, and then did it again, and again, and again. The motion a bison makes is also just like that of a fighting bull (my source on fighting bulls is Ernest Hemingway): The horn catches the victim, and the leverage the bull gains when the horn penetrates the victim's body allows him to throw a full-grown man quite literally like a rag doll. The point of a bull bison's horn is set closer to the head than that of a fighting bull, so the point can miss, but on this day, it hardly mattered. Lindsay ripped into his owner with such violence that he tore all of Rock's clothes off. One of the hired hands came to Rock's aid, drawn by his screams, and skewered Lindsay with a pitchfork, but it had no effect on the ongoing rampage. At last, a neighbor ran up and shot Lindsay. The only clothing left on Rock's body were his shirt cuffs and his socks; his trademark sombrero had been ground into the dirt. Mrs. Rock looked into her husband's face. His eyelids twitched once or twice, and then he died. It is a wonder he lasted that long.

We have not had an attack quite so relentless inside the national parks, that I know of. To date, two people have been killed by bison in Yellowstone National Park. The first death occurred on July 12, 1971, on Fountain Flats. As in so many bison stories, photography was involved. A man named Marvin Schrader approached to within twenty feet of a lone bull bison to take its photo. It was the wrong bull to approach; a group of youths had been throwing rocks at the animal, apparently trying to get it to stand up. It did, for Schrader: It charged, caught him with his horn, and threw him over a dozen feet through the air. It had delivered what Hemingway would recognize as a classic *cornada de caballo*, the worst kind of horn wound. This is "the same sort of wound in a man that the bull usually makes in the chest of a horse," as Hemingway explains in *Death in the Afternoon* (his 1932 book on bullfighting in Spain, a surprisingly useful book for understanding this topic). The bison had torn open Schrader's abdomen and penetrated his liver, and he died quickly on the ground at Fountain Flats.

It happened next in 1983. On July 31 of that year, a visitor from France, Alain Dumont, was, yes, getting a photograph of himself with a lone bull bison, in Hayden Valley near Alum Creek. He was within six feet when the animal charged, and again, the wound was a *cornada de caballo*, with his abdomen laid well open. He had four broken ribs, a mostly destroyed spleen, and a shredded stomach and colon. He died on September 2, and as with Rock, it is remarkable that he survived that long.

There has not been a death since, but it is not from lack of trying. Bison injure people almost every year, and they menace people almost every day, although "menacing" usually isn't reported to the government. The attacks happen partly because there are so many bison. They are also present in large numbers in places where people tend to cluster. They turn up regularly in the many open areas through which the Grand Loop and the five entrance roads carry the public, like Fountain Flats and Hayden Valley; there seem to always be a few hanging around near Old

Faithful Geyser, a sure recipe for trouble. They do not always flee when approached the way many animals do. They stay put, and they are such a spectacular sight that people always want a photograph. Now that everyone has both a still and video camera in his or her pocket at all times, the temptation is ever present, along with the vexing, unforgiving, *furious* need to post the images to social media.

So, the attacks happen. They have happened so often that the whole phenomenon has been written up as a threat to public health in medical literature. The Centers for Disease Control even got involved, and their people wrote it up for *Morbidity and Mortality Weekly Report*, normally the place they announce new epidemics. "Since 1980, bison have injured more pedestrian visitors to Yellowstone National Park . . . than any other animal," they reported. "After the occurrence of 33 bison-related injuries during 1983–1985 (range = 10–13/year), the park implemented successful outreach campaigns to reduce the average number of injuries to 0.8/year (range = 0–2/year) during 2010–2014." And then it all went to hell: "During May–July 2015, five injuries associated with bison encounters occurred."

The purpose of an agency like the CDC is to figure out why a sudden shift in "public health outcomes" like that happened, but know this: When government scientists are investigating events in Yellowstone, it is also partly a vacation, which is why you get research from anthropologists with titles like "Problematizing the Rituals of Scuba Diving in Aruba: A Participatory Study" (don't tell the IRS). But there were things to be learned from what happened during the rampage in 2015. What the good doctors were looking for were features the attacks shared in common, and they found them, including some we have seen already: "Every incident occurred in developed areas, such as hiking trails or geyser basins. Two persons were gored, and three were tossed into the air. Four persons required hospitalization, three of whom were transported by helicopter ambulance."

And here is a big one we have also seen. The language is that of epidemiology, with its statistics and percentages—but behold again the fatal attraction that is photography:

All encounters resulted from failure to maintain the required distance of 75 ft (23 m) from bison. Four injuries occurred when three or more persons approached the bison. Two persons were injured while walking on hiking trails. Three persons sustained injuries while taking photographs at a distance of approximately 3–6 ft (1–2 m) from bison, including two who turned their back on the bison to take the photograph; one person reported taking a cell phone self-portrait (selfie), which necessitated getting close to the animal.

During 1980–1999, a total of 10 of 35 bison encounters (29%) involved photography; the majority of persons were ≥10 ft (3 m) from the bison, unlike the 3–6 ft (1–2 m) reported with recent photography-related injuries. Smart phones now meet the needs of most casual photographers. Smart phones are owned by 64% of American adults, and 67% of smart phone owners report using their phone to share pictures and videos. The popularity of smart phone photography with its limited zoom capacity and social media sharing of selfies might explain why visitors disregard park regulations and approach wildlife more closely than when traditional camera technology was used. Educating visitors about wildlife behavior and the need to maintain distances of 75–300 ft (23–91 m) from wildlife for safety of persons and wildlife is critical. Injury prevention campaigns that identify and target the underlying motivations of visitors to not comply with viewing distances might prevent future injuries.

But they are already doing that, and have been for years.

When a new visitor enters the park, they get a copy of the park newspaper with a flyer, printed on yellow paper, with an alarming piece of art depicting a park visitor thrown by a bison. The original version of the artwork, which came into use in 1984, included, tellingly, a camera flying through the air next to its owner flying through the air. If you are presently in the park, the new version of that flyer is probably in your car. The newsletter itself, and the map, and the signs, and the website, and the visitor center, all are covered in warnings. The National Park Service has done what it can along these lines. The rest is up to the visitor.

Maybe what they need is to see some of the medical literature I have seen on this topic. The *Journal of Wilderness Medicine* wrote up a series of attacks as case reports, the kind of scientific article, important in medicine, in which specific injuries to specific patients are examined. When Ernest Hemingway wrote about *cornadas* and *cornadas del caballos*, he was usually talking about wounds that were fatal (although it is amazing what the old-time matadors could withstand and still return one day to the ring). The dozens of people who have been "merely" wounded over the years have been brutalized to a remarkable degree; it is difficult to imagine why we have had only two deaths, given how far a Yellowstone bison victim is from a real twenty-first-century trauma center.

One of the "merely" wounded was a fifty-seven-year-old woman thrown ten feet through the air. She landed with six broken ribs, a broken pelvis, and a collapsed lung; the horn left an impression of itself like a photographic negative in her thigh, making a conical depression two inches wide and two and a half inches deep filled with "devitalized tissue, dirt, and debris" that required two surgical procedures all by itself. Another was a perfectly innocent seventy-six-year-old woman employed by the company that ran the general stores. She was carrying a load of laundry when she was ambushed and thrown. She landed with an "avulsion cavity" exposing the muscles and other odds and ends near her pelvis. An avulsion cavity is a term you will more often hear from a dentist; it is the language they use to describe the hole left by a missing tooth. In this case, the hole was left by a missing horn, after the bison withdrew it from the cavity. Another, a forty-two-year-old woman who was charged and gored by a bison whose horn cut through her jeans. It then penetrated her thigh and tore upward into her abdomen, where it left bowels exposed and protruding. In this case, happily, the victim's husband was a physician. This is an advantage we have in Yellowstone; the visitors are often educated professionals (although the accountants and orthodontists are no help).

This kind of research is useful, in part because it gives some sense of where the actual dangers are in a place like Yellowstone, and therefore where the government ought to devote resources in order to keep people from getting themselves in trouble. A bear usually has to put some time and effort into injuring someone with the same awfulness that a bull bison can accomplish in a split second. As the *Journal of Wilderness Medicine* study recorded, under "Mechanisms of Injury," they do it like this:

> Goring or "hooking" occurred, resulting in punctures or lacerations from penetration by the bison horns. Blunt trauma was sustained when the person was butted or shoved or tossed into the air by the upward movement of the bison's head, encountering sudden deceleration on impact with the ground (or other structure).
>
> Goring injury was documented in 36 cases. Of these victims, 14 were tossed into the air after being hooked by the horn. A butting or shoving injury with blunt trauma only was recorded in 11 instances. In 19 cases, the bison "charged" the victim. . . . Multiple trauma, defined as goring injury plus significant blunt trauma (i.e., fractures), occurred in 19 cases. All of the multiple-trauma victims required hospital admission.
>
> Twenty of the 36 goring injuries were to buttock, thigh, or hip. . . . These were deep puncture lacerations, with a ripping and avulsion mechanism [the physician's word for "tearing loose"] extending from the initial site of skin contact/puncture. Bison horn goring wounds to the thigh or buttock caused cavities of 8–12 cm width by 14–20 cm depth.

That is, up to eight inches deep. I have been emphasizing penetrating wounds made by the horn, but the pile-driver effect of that bull bison head slamming into the victim is just as frightening. "Blunt trauma resulted in fractures, contusions, and abrasions. The distance into the air that the victim was thrown by the impact from the bison's head was consistently described as '10 ft.' . . . This estimate by witnesses was not actually measured in any incident, but a video recording made in summer 1991 shows a victim being airborne at least this height."

In the 1990s, when I worked in the park, there were not as many video cameras in the world. By the time I left, that 1991 video was being shown to incoming employees; in 1993, I was part of a group of location managers who, at a staff meeting at headquarters in Gardiner, decided to use the video as a training tool, a way of keeping our new employees from coming to grief. Actually, I think the decision had already been made, and we just wanted to see it, for the usual unseemly reasons—but it was the first of its kind. There was no YouTube in 1993.

There certainly is a YouTube today, and that precise video is on it, under the title "Bison Goring in Yellowstone," on the page belonging to the National Park Service. That version has been viewed 522,000 times. Again, the public has been warned. The sad fact is that we just have to live with the situation: Attacks will happen now and then. They seem to be unavoidable.

We can still try to stop them. Another medical investigation into the topic was published in the journal *One Health* in 2018. The authors looked at attacks between the years 2000 and 2015 and tried to determine what it was that people were doing to get themselves in trouble. Some of them, as we have seen repeatedly, get too close, trying to get the perfect photo. They found others, though, such as the odd tendency of attacks to happen to groups of three or more people, the opposite of what occurs with bears; in fact, the majority of attacks during those years occurred on people in groups of that size. The researchers thought (and some of their interviews backed them up) that the victims were being drawn into dangers they would otherwise have avoided by peer pressure. The people they were with, in one way or another, led them on or egged them on until the bison called a halt to it. They thought that bison feel threatened by the size of the group: the bigger the size, the bigger the threat. The animal is actually responding to what it believes to be a literal threat to its life: "Research has shown that animals perceive human disturbances as analogous to predation risks." Given how puny humans are compared to a bull bison, it seems as though the animal is overreacting, but, of course, humans *are* the gravest hazard to bison. The perceived threat would be all the greater if the human group was big enough that the animal thought it was being surrounded, its avenues of escape cut off.

Here is one cause of bison attacks the *One Health* researchers did not examine: alcohol. On April 21, 2024, very early in the year for it, two guys from Idaho Falls were driving on the West Entrance Road when one got out and kicked a bison in the leg, one suspects in order to get it to move out of the way. The bison in turn hurt him, although not badly enough that the two were unable to flee. They only got as far as West Yellowstone, where the NPS caught up with them. The one who did the kicking was charged with being under the influence of alcohol, disorderly conduct, and approaching and disturbing wildlife. The driver was arrested for driving under the influence, failing to yield to a police car, and "disturbing wildlife."

They can hand out as many flyers as they want, and it is never going to stop incidents like that.

If you are in the park and have not seen a bison yet, do not despair. As noted, you can see them in the Lamar River valley, up- and downriver from the Lamar Buffalo Ranch. As I did with my first, you can see them in Hayden Valley, between Fishing Bridge and Canyon. They also regularly turn up, conveniently, at the park's most popular destination: the Upper Geyser Basin, where they can be found in the meadows down-

Yellowstone traffic.

stream from Old Faithful. On the roads from the West Entrance to Old Faithful, they are such a routine sight that traffic is regularly halted. It is actually quite a problem; the West Entrance Road is so jammed that the traffic will be locked up for miles, often for a single bison. Again, go to Lamar. Don't let the single animal distract you.

But really, in summer, they can turn up almost anywhere. In the winter, their behavior shifts, and they now head downhill. Some head west, others north. Great masses of them move along the road between Tower Junction and Mammoth. They like that road especially, one assumes, because the snowplows keep it clear; in a deep-snow year, it is the only easy traveling they will experience. The general movement is down toward Gardiner and the Yellowstone River valley, and they are making this move because, as winter advances, snow gets deeper and food more scarce, even for these extraordinary animals whose head functions as a snowplow.

On the road from Mammoth to Norris.

As they move down the Yellowstone River valley, they are stopped at Yankee Jim Canyon. As we have seen, there is a cattle guard just before the pass at Yankee Jim that, together with the geology—that is, the rocky ridgeline, impossible for a bison to climb—keeps them away from the beef cattle down the valley. The highway department covers the cattle guard during the summer, because the steel bars are, when wet, a serious hazard for motorcycles, and furthermore, covering the guard does no other harm. It is not needed. The bison have plenty to eat on the other side, upstream and back inside the national park. They have to be kept from passing through this barrier because of a disease you most likely have never heard of.

The disease is brucellosis, and you have not heard of it because people in the United States do not usually get it. Humans in this country are brucellosis-free because their food is. Brucellosis is caused by a genus of bacteria, *Brucella*, that includes a number of species. The main villain, in Yellowstone, is *Brucella abortus*. The disease can be serious, even fatal, and at least a half-million people around the world get it every year, but the threat, in the United States, is mainly to cattle. *Brucella abortus*, as the name implies, causes domestic cows to abort their young, and also causes infertility. The Montana beef cattle herd—all those cows you see, coming and going from the North Entrance to Yellowstone National Park—is certified brucellosis-free, "certified" because the government says it is so. Because the government says that, and does actually check animals to make sure the claim is true, Montana ranchers can send beef anywhere in the United States. If the animals of the state had *Brucella* in them, that could no longer be allowed.

The bison and elk of Yellowstone share *Brucella abortus* among themselves freely. The cattle downstream, in Paradise Valley and beyond, have to be protected from the Yellowstone bison. They presumably also have to be protected from the elk, but that is one of the largely unasked questions in this whole affair. The bison, and their *Brucella abortus*, are what everyone is worried about, perhaps partly because they *can* worry about the bison. The elk will not be stopped by that paltry barrier in Yankee Jim Canyon.

A bull that did not make it through the winter. It is deep in the backcountry, and it is still there, in part because it is so *heavy*.

So, the bison in the northern part of Yellowstone National Park are stuck, like water behind a clog. Regularly, in the winter, they come down the hill from the upper elevations and through Gardiner. Ask the locals, and they will show you photos of bison passing through the Roosevelt Arch, bison camped out on the school athletic field, bison literally walking down the middle of US 89, bison plopped on the front lawn. They keep going until they no longer can, at which point they back up like that water behind the clog.

Another unasked question is whether the clog is the problem so many people think it must be. The assumption is that upstream from Yankee Jim Canyon, the bison will eat everything but the dirt, and will then picturesquely, problematically die. We do not have the space to consider this issue, because the controversy about overgrazing in the park has been going on unabated almost since the park has existed, and no agreement on the issues is likely. Since people feel something has to be done with the bison, something has.

If you are, again, going in and out of the North Entrance, you may have noticed bison in enclosures down the valley. They are in quarantine. So are their relatives far to the northeast, at the Fort Peck Indian Reservation, near the Canadian border. Since 2019, Yellowstone bison have been trucked there, where their quarantine continues. They are tested there, and in the Yellowstone pens, for brucellosis, and repeatedly. If they should prove to be healthy, they can be shipped elsewhere. Another issue comes into play here, not one of bacteriology, but genetics. Bison have, for decades, been extensively crossbred with beef cows, to come up with a new kind of domestic steak-and-hamburger-bearing organism. The resulting animal has been called a beefalo. If you had a buffalo burger in a restaurant in Greater Yellowstone, that meat came from such an animal. The Yellowstone bison have never been crossed in this way. There is some debate about their genetics (they seem to be a mix of two different subspecies of bison, wood bison and plains bison), but we know for sure that they have never been crossbred.

As of the date I am writing, a total of 414 Yellowstone bison have been handed over to the Assiniboine and Sioux Tribes at Fort Peck. Most of them have been handed off, in turn, to twenty-six other tribes. The goal is to transfer bison to tribal land all over North America.

That number, 414, may look impressive—but if you think about it, you will see the problem. There are between 3,000 and 6,000 bison in Yellowstone National Park at any given time; at present, there are 4,550. If you take my recommendation and visit the Lamar River country, especially in the spring, you will see a great many calves. This species reproduces fast. How to get rid of the "surplus," assuming for the moment that it really has to be done?

The solution the government has come up with—various agencies are involved—has been a brilliant one, because it is almost immune to criticism: American Indian tribes are allowed to hunt the bison. There is plenty of criticism anyway, but it has never stuck. The hunt continues. Every year, the bison descend, and so do the tribes, who bring a small rush of business to the merchants of Gardiner. The tribal hunters go out and shoot a bison, or an elk; the hunt is conducted as an expression of the rights the tribes have from the treaties they signed, which "may mean a specific hunt for bison that includes opportunistic taking of other animals, like elk," in the language of the lawyers. One other force descends on Gardiner: reporters, who get a quote from the hunters that sounds like this, from the April 4, 2023, issue of the *New York Times*: "It's a very cultural and spiritual endeavor and brings our families together," one of the tribal hunters said. "And it gives us an opportunity to talk about who we are and where we come from." Another shot a bull, and the reporter described the scene like this: "After it fell to the snow, she and her husband gutted it, and she took a ceremonial bite of the old bull's heart. 'That is a sign of respect,' she explained. 'Everything we carry is within our heart. A big bull like that has made it through all the different seasons and territorial fights with other bulls, and you are taking on its spirit and the different teachings it has within it.'" The hunters have Anglo-American first names—like Tom, Dick, and Harry—but last names like Wind-In-His-Hair and Stands-With-a-Fist. That is the genius of it. Under these circumstances, the government can hunt the bison to extinction, and no one can object, because to do so would be a hate crime.

But people do object, because it is an ugly scene. It is not a hunt. The shooting is done, a lot of it, at a place called Beatty Gulch, just over the boundary between Yellowstone National Park and the Custer Gallatin National Forest. The bison wander over the line, and a tribal hunter blasts it. There is no "hunting" involved. Further, some of the hunters are skilled, but inevitably not all. Too many people are involved, and so we get hunters who spend most of the year, as the rest of us do, doing their hunt-

The positive side of the tribal hunt: a trailer loaded with wild meat, headed home.

ing and gathering at McDonald's. Their aim is not always the best.

During winter 2023, people did object, loudly. That was a brutal winter, cold and lengthy. Bison came out of the park in large numbers. The hunt ended up "taking" 1,150 bison, a big chunk of that precious herd. I missed most of that hunt myself, but the next spring, I hiked along the edge of Beattie Gulch, headed down after a climb up the north flank of Electric Peak. I found myself on a hill just above a wide flat that extends away from the Old Yellowstone Trail. Here is where much of the tribal hunt happens.

I was looking, actually, for spent ammunition. I did not actually find any that day—the hunters had done the right thing and cleaned up the brass. Then I looked again, across the flat, and noticed them: the picked-over remains of bison carcasses, dozens and dozens of them. They were everywhere.

And I laughed. It was in poor taste, but I was alone, and got away with it. I laughed because the scene reminded me of another, and the irony was just too much to withstand.

One of the victims, in Beattie Gulch.

A clear miss: a bullet hole in the animal's hump.

It reminded me of the pictures you see of bison slaughtered in the 1880s.

Bodies everywhere, animals having fallen in a way that re-created, in death, the herd they had been a part of. The scene in Beattie Gulch was a mirror image of the living herd, now made only of rapidly bleaching bone and hide. The scene looked just like one of those photos of bison slaughtered.

By white hunters.

No one wants to crash the beef economy of Montana (well, some hippies I know, maybe). This whole affair would be enormously easier if it were not for the damnable *Brucella* (appropriate that the genus of bacteria has a name that makes it sound like a Halloween witch).

This is, at least, not the end of their story. Unless something goes really amazingly wrong, the Yellowstone bison are still saved. So is that precious genome wrapped up inside their hides. Scientists will be studying them for generations. Let us hope they remember to keep their distance.

Conclusion: Be Not Afraid

Whenever I write about Yellowstone, I come back to a few basic ideas and attitudes that have served me well since that afternoon in May 1990 when I first drove up US 89 and into Yellowstone National Park, leading to that meeting with the bull bison, along with a whole lot of other stuff. That first series of summers was partly a matter of learning these attitudes. I picked them up from authors and books that I discovered along the way, often entirely by accident. The most improbable was a short book published in 1942 by José Ortega y Gasset, *Meditations on Hunting*. Ortega was Spain's best-known academic philosopher of the twentieth century. The title is accurate: The book is a contemplation of hunting, by a hunter, and is regularly quoted in brief snippets in hook and bullet magazines.

This all needs explaining. For Ortega, hunting has nothing to do with bloodlust, and is much more than a diversion. Hunting "alone permits us the greatest luxury of all, the ability to enjoy a vacation from the human condition through an authentic 'immersion in Nature.' But that immersion is not as easy to achieve as is usually supposed without thinking about it. Man cannot re-enter Nature except by temporarily rehabilitating that part of himself which is still an animal. And this, in turn, can be achieved only by placing himself in relation to another animal." That sounds weird, until he explains that, once in the field, "the hunter begins to behave like the game. He will instinctively shrink from being seen; he will avoid all noise while traveling; he will perceive all his surroundings from the point of view of the animal, with the animal's peculiar attention to detail. This is what I call being *within* the countryside. Only when we see it through the drama that unfolds in the hunt can we absorb its particular richness. . . . Wind, light, temperature, ground contour, minerals, vegetation, all play a part; they are not simply there, as they are for the tourist or the botanist, but rather they *function*, they act."

This heightened awareness puts the hunter in a unique position: "The tourist sees broadly the great spaces, but his gaze glides, it seizes nothing, it does not perceive the role of each ingredient in the dynamic architecture of the countryside. Only the hunter, imitating the perpetual alertness of the wild animal, for whom everything is danger, sees everything and sees each thing functioning as facility or difficulty, as risk or protection."

Ortega sums up the hunter's mental condition in a single word: "alertness." While in the field, the hunter is the exact opposite of "passive," is instead totally alive in every way, all senses taut and working as they were meant to, and the mind also. The human body has not changed much in the last ten thousand years, and in hunting it gets the drill it was built for; in a world where people devote immense energy to fads, here is a pursuit that is solid, honest, real. Hunting provides an intensity of experience that is available in only a few other pursuits. One of these pursuits is traveling through Greater Yellowstone, on foot.

In hunting, you are trying to outwit the animal. In hiking in a place like Yellowstone, you are trying to outwit the animal to avoid being eaten. All through this book, I have mostly just implied the real reason for dwelling on the dangers of Greater Yellowstone. It is this: The danger provides that immersion into the countryside that Ortega found so enchanting. Without the danger, you are just kind of goofing off. With it, you have everything; because you are surrounded by potential threats, you are fully alive. I do not even think it requires any extreme exertion. Merely being on foot in the backcountry can be riveting.

And yet you are supposed to do everything in your power to nullify it. While I was doing the research that went into the bulk of this book, I was continually irritated by the commentary I found online, especially on social media. There is almost a genre of national park analysis in which writers of social media posts are at pains to tell us how very smart they are, compared to someone in the news who has just been killed or injured. A hiker gets attacked by a bear, and Reddit responds: "How *stupid!*" A park visitor gets kicked by an elk, and Facebook responds: "What an *idiot!*" A park employee gets sledgehammered by a bison, and Instagram responds: "What an absolute *moron!*" All along I have thought that the people making these comments were the more sedentary users of those platforms. They also, perhaps, understand that life is getting away from them, and to crush the feeling of defeat and foreboding, they have a jolly time denouncing people who tried to get the most from life and got kicked or mauled for their effort. And I got annoyed in part because by the standards of Reddit, Facebook, Instagram, and all the others, the single *stupidest, most moronic moron of all* was . . . John Muir.

He is the one historical personage whose face you are most likely to see in a national park, looking down from the wall of a visitor center or off the cover of a book at the bookstore: John Muir (1838–1914), the transplanted Scot who did much to found both the national park system and political environmentalism itself. He was also, by the way, one of the most accomplished scientists of his generation, and in multiple fields. The gentle visage and praiseworthy life of this extraordinary man is a heck of a place to find the *stupidest stupidity*, but consider how he traveled in the wilderness.

Nowadays, we are informed, always, that the only way to travel in bear country is in a group of no fewer than four. That is a great idea if you can arrange it, but Muir could never have found four people who could keep up with him. He would have had to stop hiking permanently if he insisted on traveling in such a gigantic group; instead, he literally never hiked any other way except alone, and he never gave the impression that he thought he was tempting fate.

One of the more hair-raising moments in Muir's life was the time he clambered out onto the verge of Yosemite Falls. Here is a description by the historian Bill Youngs:

> *One fine summer day in 1868 the young John Muir stood by a cliff edge hung over the Yosemite Valley. A half mile below, the land "seemed to be dressed like a garden—sunny meadows here and there, and groves of Pine and Oak." Nearby the Yosemite Creek cascaded through a channel in the rock, sped down a short incline, and sprang "out free in the air." Muir wanted a clear sight of the waterfall and began to work his way down the rock. Below he could see a narrow*

shelf that might support his heels over the sheer cliff. He filled his mouth with artemisia leaves, hoping the bitter taste would prevent giddiness. He then worked his way along the ledge; it held him, and he was able to shuffle twenty or thirty feet to the side of the falls. There he found what he wanted—"a perfectly free view down into the heart of the snowy, chanting throng of comet-like streamers, into which the body of the fall soon separates." In camp after dark that night he recorded in his journal that "the tremendous grandeur of the fall" had smothered all fear. He had had a "glorious time."

There was, beneath his feet, about fourteen hundred feet of sheer empty space.

Or how about the famous occasion, in 1874, when he climbed to the top of a tree in an unusually violent storm? Here is Muir's own description, from *The Mountains of California*:

After cautiously casting about, I made choice of the tallest of a group of Douglas Spruces that were growing close together like a tuft of grass, no one of which seemed likely to fall unless all the rest fell with it. Though comparatively young, they were about 100 feet high, and their lithe, brushy tops were rocking and swirling in wild ecstasy. Being accustomed to climb trees in making botanical studies, I experienced no difficulty in reaching the top of this one, and never before did I enjoy so noble an exhilaration of motion. The slender tops fairly flapped and swished in the passionate torrent, bending and swirling backward and forward, round and round, tracing indescribable combinations of vertical and horizontal curves, while I clung with muscles firm braced, like a bobo-link on a reed.

In its widest sweeps my tree-top described an arc of from twenty to thirty degrees, but I felt sure of its elastic temper, having seen others of the same species still more severely tried—bent almost to the ground indeed, in heavy snows—without breaking a fiber. I was therefore safe, and free to take the wind into my pulses and enjoy the excited forest from my superb outlook.

Conclusion

Or how about the time that the snow slipped out from under him and he literally rode an avalanche to the bottom of a mountain? He could not have been more delighted. Again, his description, from "The Fountains and Streams of the Yosemite National Park":

Few mountaineers go far enough, during the snowy months, to see many avalanches, and fewer still know the thrilling exhilaration of riding on them. In all my wild mountaineering I have enjoyed only one avalanche ride; and the start was so sudden, and the end came so soon, I thought but little of the danger that goes with this sort of travel, though one thinks fast at such times. One calm, bright morning in Yosemite, after a hearty storm had given three or four feet of fresh snow to the mountains, being eager to see as many avalanches as possible, and gain wide views of the peaks and forests arrayed in their new robes, before the sunshine had time to change or rearrange them, I set out early to climb by a side cañon to the top of a commanding ridge a little over three thousand feet above the valley. On account of the looseness of the snow that blocked the cañon I knew the climb would be trying, and estimated it might require three or four hours. But it proved far more difficult than I had foreseen. Most of the way I sank waist-deep, in some places almost out of sight; and after spending the day to within half an hour of sundown in this loose, baffling snow work, I was still several hundred feet below the summit. Then my hopes were reduced to getting up in time for the sunset, and a quick, sparkling home-going beneath the stars. But I was not to get top views of any sort that day; for deep trampling near the cañon head, where the snow was strained, started an avalanche, and I was swished back down to the foot of the cañon as if by enchantment. The plodding, wallowing ascent of about a mile had taken all day, the undoing descent perhaps a minute. When the snow suddenly gave way, I instinctively threw myself on my back and spread my arms, to try to keep from sinking. Fortunately, though the grade of the cañon was steep, it was not interrupted by step levels or precipices big enough to cause outbounding or free plunging. On no part of the rush was I buried. I was only moderately imbedded on the surface or a little below it, and covered with a hissing back-streaming veil of dusty snow particles; and as the whole mass beneath or about me joined in the flight I felt no friction, though tossed here and there, and lurched from side to side. And when the torrent swedged and came to rest, I found myself on the top of the crumpled pile, without a single bruise or scar. Hawthorne says that steam has spiritualized travel, notwithstanding the smoke, friction, smells, and clatter of boat and rail riding. This flight in a milky way of snow flowers was the most spiritual of all my travels; and, after many years, the mere thought of it is still an exhilaration.

Look what happened there. He set out to climb up a wall of highly unstable snow specifically *because* it was unstable. And in his era, there were no space blankets, no water purification tablets, no aluminum ice axes, no high-tech crampons, no cell phones to use to call the National Park Service for help—indeed, no National Park Service.

I could go on; indeed, I could fill another book the size of this one with just such incidents, described by his own pen. Muir's life was filled with episodes like these, not always as death-defying as, say, his scramble around the brink of Yosemite Falls (which gives me the absolute creeps), but routinely exciting.

We can say a few things for sure about him. He was a highly skilled woodsman. He was a little lucky. He also lived life to the fullest, in the manner José Ortega y Gasset recommends, even though Muir was no hunter. He lived, and he was rarely bored.

Nor was he beset by the fake life, as we are today. I am not recommending that you go climbing trees in violent storms, or riding avalanches, or gripping the edge of a granite dome in Yosemite by your toes. But you do not necessarily need to. Danger is a matter of degree, so that, for example, people who in some way face physical challenges (including older people) can have a perfectly thrilling time without going too far. If you want to give that fake life the slip, it *is* necessary to explore Greater Yellowstone on foot, and away from the asphalt and the cars. You do not, however, have to attack the place like Edmund Hillary and Tenzing Norgay assaulting Mount Everest. The goal is to see Greater Yellowstone for what it is, and to get from it all you can. That it can, in fact, kill you is the precise quality that will make you come to life.

It is there for you, right now. It is always there. "The mountains are calling," Muir said in a letter to his sister, "and I must go."

So go.

Just don't try to get a selfie with your arm around a bison's shoulder. That would just be *stupid*.

www.ingramcontent.com/pod-product-compliance
Ingram Content Group UK Ltd.
Pitfield, Milton Keynes, MK11 3LW, UK
UKHW041822310326
5020IPUK00002B/23